Praise for THE 7 LAWS OF MAGICAL THINKING

'In this wickedly funny and deeply clever book, Matthew Hutson makes a radical claim: All of us, whether we accept it or not, believe in magic. Without these intuitions, he says, we would hardly be human. Through vivid examples and cutting-edge science, Hutson presents a provocative new theory of how we make sense of the world.'

Paul Bloom, Professor of Psychology, Yale University, and author of *How Pleasure Works*

'This book about thinking is magical. It's the perfect blend of astonishing stories, up-to-date science, awe, beauty, disgust, and humour. It's science journalism at its best: great writing and deep humanity bring out the profound relevance of psychological experiments for people who search for meaning using minds that were designed for so many other purposes.'

Jonathan Haidt, author of *The Happiness Hypothesis*

'This is a book that you pick up, but can't put down. Hutson, intelligently and entertainingly, gives us the best kind of book, one that gives us insight to our very core.'

Ori Brafman, co-author of *Sway* and *Click*

'Brilliant, exhilarating ... Reading this book will lead you to a more reverential appreciation of human irrationality, and the science that tries to understand it.'

Dacher Keltner, Director, Greater Good Science Center, University of California, Berkeley

THE
7 LAWS
OF
MAGICAL
THINKING

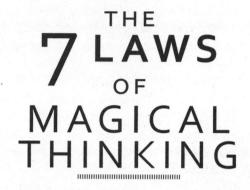

How Irrationality Makes Us
Happy, Healthy, and Sane

MATTHEW HUTSON

ONEWORLD

A Oneworld Book

First published in Great Britain and the Commonwealth
by Oneworld Publications 2012
Published by arrangement with Hudson Street Press,
a member of Penguin Group (USA), Inc
This edition published by Oneworld Publications 2013

A CIP record for this title is available
from the British Library

ISBN: 978-1-85168-957-6
Ebook ISBN: 978-1-78074-109-3

Cover design by Dan Mogford
Printed and bound by Nørhaven A/S Denmark

Oneworld Publications
10 Bloomsbury Street, London WC1B 3SR

Stay up to date with the latest books,
special offers, and exclusive content from
Oneworld with our monthly newsletter

Sign up on our website
www.oneworld-publications.com

For my teachers – past, present, and future

Matthew Hutson is a science writer and the former news editor of the magazine *Psychology Today*. His work has appeared in the *New York Times Magazine*, *Wired*, *Scientific American Mind*, *Discover*, and many others. He is an atheist and magical thinker.

CONTENTS

Introduction

We're All Believers

In 2008, the leaders of a powerful clan presided over a ceremony on the grounds of their new house of worship. The clan's warriors, known for their fickleness and inconsistency – their success against other tribes depending to a large degree on luck – worried that an adversary had placed a curse on their home turf. Someone had hidden a significant artefact – a symbol of their sworn enemy – under the premises. The media, typically dismissive of voodoo, had a field day with this little rite. As journalists looked on, two men friendly to the warriors pulled the offending relic from the ground and raised it high. Flashbulbs illuminated a ragged piece of cloth clearly reading the number 34 and the name Ortiz. The new Yankee Stadium had been cleansed.

Why should an enlightened society adhering to the rigours of science care so much about a shirt buried in concrete? And why would the president of the New York Yankees baseball team threaten the offender with legal action and demand recompense for the cost of replacing the concrete? The jersey – carrying the number and name of David Ortiz, the top home-run hitter for the rival Boston Red Sox – itself posed no structural threat to the stadium. So how could that worker 'force' the Yankees to dig it up? Because magical powers were attributed to that jersey. (We'll revisit Yankee Stadium in chapter 2.)

Most of the world is religious, and millions more are openly super-stitious, spiritual, or credulous of the paranormal. But in this book I argue that we all believe in magic – luck, mind over matter, destiny, jinxes, life after death, evil, and heavenly helpers – even when we are sure we don't.

Magical thinking can be quite banal. We find occult meaning in the world all around us, every day. Do you own any sentimental objects – say, a wedding ring, a family heirloom, or an autographed football shirt? Objects you'd value more than an identical duplicate? That's magical thinking. Do you feel that what goes around comes around, through some universal principle of fairness? That's magical thinking. Do you shout at your laptop when it erases your files? Magical thinking. Do you hope to leave a legacy after you die? Magical thinking. Do you believe that certain events were meant to happen? Magical thinking. Or that you can lift your arm through the power of your conscious thoughts? Magical thinking, even that.

As you will see, those examples all derive from our ongoing flirtation with supernaturalism, a relationship we depend on for our very survival.

Giving Up the Ghost

For the first ten years of my life I went to church every week with my family. Not by choice; I found it boring and hated getting up early and wearing uncomfortable clothes. But we got doughnuts in Sunday school, I enjoyed a modest version of stardom as a member of the choir, and I was allowed to spend sermons drawing tanks and fighter planes blowing up the illustration of the church on the cover of the programme.

And I did believe in, and fear, God. I hated being alone with him in the empty chapel – it gave me goose pimples. For a time I refused to say the word *God* and would spell it out. I even wrote it 'G-O-D'.

But things changed when I was about ten years old, when I discovered a copy of *A Brief History of Time* by Stephen Hawking on my parents' bedroom floor. I read his portrayal of the evolution of the universe, first with my father and then on my own, and saw that the Big Questions could be answered, or at least approached, by science. God made less and less sense.

I found more books on the big bang and the fabric of space-time and abandoned my belief in a personal creator – but not my obsession with him. I became a strident young atheist, eager to debate anyone who stooped to have faith in an invisible guide. In the copy of *Why I Am Not a Christian* by the philosopher Bertrand Russell that I purchased for pleasure-reading when I was about twelve, I underlined passages such as, 'It would seem, therefore, that the three human impulses embodied in religion are fear, conceit, and hatred'. I struggled to understand human-ity's unshakeable hold on magical beliefs – its stock in miracles, gods, a soul – against all reason.

That's just it: faith is unreasonable, an emotional reaction. But shouldn't reason triumph in deciphering the workings of the universe? Why cry out for a daddy in the sky to explain things and keep you safe? (I have Freudian interpretations of my conversion, too, but I'll save those for psychotherapy.) In my Vulcan mind-set, I looked down on the religious as stupid or weak or both.

But I knew too many intelligent, admirable people who went to church. Besides, I never converted anyone to atheism using logic. So I decided to relax and pay more attention to what irrational beliefs did for people. Five billion faithful can't be wrong!

And I realized in myself a continued need for something more. My teen years were dark, and I often thought that life would be easier were I not an atheist. I looked for slivers of evidence to let me believe that we are not simply mortal, finite, arbitrary collections of organic molecules. I read *Synchronicity*, in which the physicist F. David Peat tries to ground Carl Jung's ideas about meaningful coincidences in the world of quantum mechanics. I read *The Physics of Immortality*, in which the physicist Frank Tipler proposes that our descendants will use com-puters to re-create all previous humans and continue our existences in a virtual heaven. And I read *The Archaic Revival*, in which the ethno-botanist Terence McKenna considers psychedelics a window into higher dimensions. (Naturally, I also tested some of those windows.) And here is something I've never told anyone before. For a couple years after giving up God, I still occasionally prayed at night, sending my thoughts out into the vast ether.

I really, really wanted to believe in magic.

In parallel with my search for meaning was the pursuit of the *meaning* of meaning, which led me from physics to psychology. We can't interact with reality directly and in fact can't even be sure it exists; we experience it only through the filter of our own consciousness. What you see, hear, taste, and touch is all a subjective construction in your brain based on sensory input. (Or a neural jack, as in *The Matrix*.) I decided the closest I could come to understanding the ultimate nature of reality was to understand how the mind creates it. At university I set out to design an independent course in consciousness studies before settling on cognitive neuroscience, the rigorous analysis of the interface between matter and mind, existence and experience.

That pursuit has led me here. I can't of course provide for you the meaning of life, and might even speak dismissively (though not derisively) of the meaning you already hold dear. But I'm not ruining Christmas just for fun. (And, arguably, I'm not ruining Christmas at all; telling people why they're biased to believe in Rudolph says nothing about Rudolph's actual existence.) I'm dissecting the sacred because the same magical thinking that leads to sentimentality, altruism, and self-efficacy can also lead to vilification, fatalism, and irrational exuberance, or even depression, obsessive-compulsive disorder, and psychosis. By tearing down everything holy and pointing out the sand it was built on, I'm hoping we can learn how to build meaning back up in constructive ways. I don't want to eradicate magical thinking. I want to harness it.

The Rationality of Irrationality

Far from a sign of stupidity or weakness, magical thinking exemplifies many of the habits of mind that made humans so evolutionarily successful. Once you've accepted that the brain constructs reality, and that the brain has evolved like any other organ to help its owner survive and reproduce, it follows that the brain constructs reality in the most useful way possible for its owner. The key word here is *useful*, which is not to say *accurate*. The brain doesn't care so much what's really out there; it just needs to stay alive and be replicated, which might involve telling us a white lie now and again.

Over the past several decades, psychologists have documented a litany of cognitive biases – consistent misperceptions of the world – and explained their positive functions. For example, we overestimate heights when looking down, making us particularly cautious about falling. In the social realm, men overestimate sexual interest from women because the cost of hitting on someone and receiving a brush-off is small compared to the benefit of scoring and spreading one's seed. (A drink in the face is temporary, but a carrier for your genes lasts generations.) And superstitious rituals such as crossing fingers may result from believing we have more control over the world than we actually do, a bias that prevents counterproductive feelings of helplessness.

The behavioural economist Dan Ariely, who has designed many clever studies to tease out our biases, calls the human mind 'predictably irrational'. Alternatively, the evolutionary psychologist Martie Haselton and her colleagues have written that 'the mind is best described as *adaptively rational* . . . equipped with mechanisms that are constrained and sometimes imprecise, but nevertheless clear products of natural selection showing evidence of good design'.

This design comprises two distinct levels of processing. The *rational* system is slow, deliberate, abstract, and logical. The *intuitive* system is quick, automatic, associative, and emotional. We have the second system to thank for magical thinking.

Thinking and *belief*, as I use the terms in this book, include biases and intimations and feelings. Mere whiffs and glimmers of thought. If you think conscious deliberation drives the car, you're ignoring the vast engine block beneath the hood at your own peril. We run largely on autopilot, and overthinking things (as I and many others are wont to do) can muck up the works. For example, when an injury disconnects emotional brain centres from neural areas responsible for higher cognition, patients can't listen to their guts and have trouble making even simple decisions. Recall the millipede who was asked how he knows which leg to move next and immediately froze. Sometimes intuitive thinking just gets the job done. And as we'll see, magical thinking is not merely an eccentric extension of healthy biases and shortcuts; it can provide benefits of its own. Most prominently, it offers a sense of control and a sense of meaning, making life richer, more comprehensible, and less scary.

Often, the biologically modern deliberative system is powerless to restrain the ancient associative system it's built on. It makes no difference how clever you are or how reasonable you try to be: research shows little correlation between people's levels of rationality or intelligence and their susceptibility to magical thinking. I 'know' knocking on wood has no mystical power. But my instincts tell me to do it anyway, just in case, and I do. A possibly apocryphal tale has the legendary physicist Niels Bohr responding to a friend's inquiry about the horseshoe he'd hung above his door: 'Oh, I don't believe in it. But I am told it works even if you don't.' 'There are many layers of belief', the psychologist Carol Nemeroff, who has studied magical thinking extensively, told me. 'And the answer for many people, especially with regard to magic, is, "Most of me doesn't believe, but some of me does".'

Longings and Wisdom

'Magic – the very word seems to reveal a world of mysterious and unexpected possibilities!' the Polish anthropologist Bronisław Malinowski wrote in 1925. 'Even for the clear scientific mind the subject of magic has a special attraction. Partly perhaps because we hope to find in it the quintessence of primitive man's longings and of his wisdom – and that, whatever it might be, is worth knowing. Partly because "magic" seems to stir up in everyone some hidden mental forces, some lingering hopes in the miraculous, some dormant beliefs in man's mysterious possibilities.'

Malinowski spent several years in the southwest Pacific studying the magical practices of 'primitive man'. Much of today's scholarship on magic derives from the anthropological efforts of the late nineteenth and early twentieth centuries, in which the traditions of bushmen and remote islanders were catalogued and scrutinized. Psychologists, sociologists, and historians have still not agreed upon what counts as magic, versus religion, versus science, versus technology. There's plenty of overlap: magic and religion both deal with a spiritual realm. Magic and science both deal with uncovering hidden patterns in the world. And magic and technology both deal with mastering one's environment.

'Although the word "magic" is common in both scholarly and lay discourse', the psychologists Carol Nemeroff and Paul Rozin have written:

> the variety of things to which it refers is far-reaching, ranging from a social institution characteristic of traditional societies, to sleight-of-hand or parlor tricks, to belief in unconventional phenomena such as UFOs and ESP, to sloppy thinking or false beliefs, and even to a state of romance, wonder, or the mysterious. One must at least entertain the possibility that there is no true category here at all. Instead, the term 'magic' in current usage has become a label for a residual category – a garbage bin filled with various odds and ends that we do not otherwise know what to do with.

There *is* a common thread that holds together many of the things we tend to call *magic* and excludes many of the things we don't. One recurring theme in the literature – a theme I'm taking as the basis for my definition of *magical thinking* – is what the anthropologist Richard Shweder called a 'confusion of subjectivity and objectivity' and the anthropologist Claude Lévi-Strauss called 'the anthropomorphism of nature . . . and the physiomorphism of man'. There's the world of the mind, defined by intention and conscious experience, and the world of outside reality, defined by matter and deterministic forces. But we instinctively treat the mind as though it had physical properties, and we treat the physical world as though it had mental properties. That's magical thinking. We perceive mind and matter mingling together, working on the same wavelength.

The psychologist Marjaana Lindeman similarly defines magical thinking as 'category mistakes where the core attributes of mental, physical, and biological entities and processes are confused with each other' and has collected evidence linking these category mistakes under one umbrella. She and collaborators found that people who describe phrases such as, *Old furniture knows things about the past*, or, *An evil thought is contaminated*, or *When summer is warm, flowers want to bloom* as more than metaphor also believe in feng shui and astrology (i.e., that the arrangement of furniture or stars can channel life energy), see more purpose in natural and random events, and are more likely to be religious and hold paranormal beliefs.

One advantage of defining magical thinking as the mingling of psychological concepts with physical ones, rather than simply as holding beliefs that contradict scientific consensus, is that what counts as magical thinking is less prone to change as we learn more about the world. We now know that our planet is a sphere, but learning that it has a personality would constitute a revolution an order of magnitude larger.

The current definition also distinguishes magical thinking from everyday false beliefs such as the notion that toilets tend to flush clockwise in the Southern Hemisphere or that toilet seats transmit HIV, from common biases and states of mind such as germ phobia and wishful thinking, and from credence in possible but unlikely phenomena such as Bigfoot's existence and alien authorship of crop circles.

The Agenda

With our promiscuous mixing of the mental and physical realms, it's hard to break magical thinking into distinct laws, but I've tried. Someone else might divide the material differently to what I have done, with different laws, or more laws, or fewer. And things I call magical thinking someone else might dismiss as run-of-the-mill irrationality. Surely, I'll also make what some consider omissions. I've tried to take a consistent approach in mapping the terrain, but the borders remain debatable. Here's a rough guide:

In chapter 1, 'Objects Carry Essences', we'll explore how everyday items become emotionally significant by taking on the spirit of their previous owners or unique pasts. In chapter 2, 'Symbols Have Power', we'll see that we confuse symbolic associations in our heads for causal relationships in the world. Chapter 3, 'Actions Have Distant Consequences', takes up superstitious rituals and our attempts to channel luck through physical acts. Chapter 4, 'The Mind Knows No Bounds', covers belief in mind over matter and extrasensory perception, as well as transcendent experiences. In chapter 5, 'The Soul Lives On', we'll look at how hard it is to believe that your mind dies when your body does. In chapter 6, 'The World Is Alive', we'll see that we often treat inanimate objects as conscious. Chapter 7, 'Everything Happens for a Reason', analyses our insistence that higher powers guide natural events. Finally, the epilogue explores ways to find meaning in life by treating the world as sacred.

For the most part I don't cover explicit and culturally transmitted beliefs in religion, magic, and the paranormal. Plenty of excellent books exist on those. I'm more interested in our shadow beliefs – those inklings of the numinous that we deny – and beliefs we don't even recognize as magical. These habits of mind guide us through the world every day. In very basic ways they provide a sense of control, of purpose, of connection, and of meaning, and without them we couldn't function. So here's my gauntlet: even if you're a hard-core sceptic who walks under ladders and pronounces 'New Age' like 'sewage', you believe in magic.

And that's nothing to be ashamed of.

1

||||||||||||||||||||||||||

Objects Carry Essences

'Cooties', Contagion, and Historicity

Years after his death, John Lennon went on tour. He visited, among other locations, Oklahoma City, New Orleans, and Virginia Tech University, spreading a message of peace and love at the sites of tragic events. You may not have recognized him, though, covered in scars and cigarette burns. But to hear him, there would have been no mistaking his presence.

On this journey, Lennon assumed the form of a piano, specifically the one on which he composed 'Imagine'. 'It gives off his spirit, and what he believed in, and what he preached for many years', says Caroline True, the tour director and a colleague of the Steinway's current owner, the singer George Michael. Free of velvet ropes, it could be touched or played by anyone. According to Libra LaGrone, whose home was destroyed by Hurricane Katrina, being with the piano 'was like sleeping in your grandpa's sweatshirt at night. Familiar, beautiful, and personal.'

LaGrone didn't know who John Lennon was at age five but maintains a flashbulb memory of his death on 8 December 1980. 'We were decorating our Christmas tree the day he was shot. I remember I was standing on a little stepladder, putting tinsel on the tree, and my mom sat down in front of the television and started to cry. I have a very, very close family – my mom is truly my best friend in the world, and has been my whole life – so to see my mom sit down and cry over someone, it was like, "Okay, who is this guy? I'm gonna love him, aren't I?" Growing up he became a really important part of my life.'

On Friday, 26 August 2005, LaGrone helped her ex-boyfriend move his stuff out of her house in New Orleans, Louisiana. They'd been dating for six years and had broken up three days earlier. She was also caring for a cat who'd undergone intensive surgery for cancer. Understandably, she went out on the town that night to see a band with some friends and blow off steam. After returning home in the wee hours, she woke up at 8 A.M. to an incessantly ringing phone. Everyone was calling to tell her to get out of the city; a hurricane was coming. 'I put my cats, dogs, me, a pair of jeans, a pair of flip-flops, and a Clash T-shirt in my car and drove to my aunt and uncle's house in Baton Rouge', she says. LaGrone would live there, about eighty miles from New Orleans, for several months. 'I lost everything.'

Less than a month after Katrina came to town, Hurricane Rita paid a visit, and LaGrone's grandparents also evacuated to LaGrone's aunt and uncle's house. Her grandmother had a stroke. After an extended stay in the ICU, they let her pass. LaGrone's grandfather followed soon after. While her mother dealt with the aftermath, her father slipped, hit his head, and suffered a brain haemorrhage. Then both of LaGrone's cats died. And she was held up at gunpoint. It was a rough year and a half.

You can imagine the look on LaGrone's face in May 2007 when a woman showed up to the Ogden Museum of Southern Art, where LaGrone worked, with a strange offer.

'When she approached me, she said, "Hey, I'm Caroline True. I'm taking John Lennon's piano around the country to different places that have been struck by devastation".' True wanted to set the piano up at the museum so people could interact with it. 'Well, I didn't really believe her at first. I was like, "Okay, you're here to tell me you have the piano that John Lennon wrote the song 'Imagine' on in a truck in front of the museum right now? Sure, whatever, sweetheart".' LaGrone ran upstairs, Googled True, and realized she was the real deal.

How would you react if someone showed up at your doorstep with a personal belonging of one of your heroes? Chances are you'd get a little weak-kneed or at least become intrigued. You'd have to be pretty cold to say, 'So what? I've seen a piano before.'

We all have objects we fetishize or at least take a fancy to for personal reasons. Think of a child's comfort blanky. A wedding ring. An autographed album. What's more human than sentimentality? And yet, sentimentality for particular objects is just as illogical as belief in ghosts. Nevertheless, you wouldn't be happy with replicas of sacred items, even if the replica was identical in every way. Somehow two things can be physically identical, and yet the original has *something else:* a nonphysical presence, a special significance beyond the realm of physics or chemistry. How is it that an object, such as Lennon's Steinway, can carry meaning around with it, independent of its material composition? To attribute personal value – a subjective property – to a lump of atoms: that's magical thinking.

Historicity

Like most forms of magical thinking, feeling the power of unique objects appears to come naturally. When scientists want to understand why people think the way they do – particularly when they want to separate nature from nurture – they often look to young children. If you're trying to grasp the evolutionary origins of human thought and behaviour, children – relatively untouched by modern cultural indoctrination – are the next best things to cavemen. So when a scientist wants to know what types of ideas people naturally accommodate – or reject as impossible or inane – he will sometimes try telling tall tales to pristine minds, and see if they'll buy it.

What kind of a scientist lies to a child? Well, the developmental psychologists Bruce Hood of the University of Bristol and Paul Bloom of Yale are two of the most practised kiddie conmen out there. And they have one clever set of experiments that serves our case here very well. First they rigged up a 'copying machine' capable of 'duplicating' objects placed inside – really just two boxes spruced up with lights and buzzers and placed in front of a black curtain. Then they invited parents to bring their three- to six-year-olds into the lab along with one of their child's toys, dolls, or blankets. Half the objects were considered 'attachment objects', meaning the children were especially possessive of them and used them for self-soothing.

To demonstrate their machine's powers, Hood and Bloom placed various items in one box, closed the lid, snuck a duplicate into the other box from behind, then opened them both up. None of the children accused the scientists of shenanigans. The children were then invited to place their own item in the machine and asked whether they would want to take home the original or the copy. Of the children with non-attachment items, 62 percent chose the duplicate. Of the children with attachment items, only 23 percent chose the copy, and 20 percent refused to put their beloved toy in the machine at all.

Are children, like adults, suckers for the celeb worship by which objects can be important just because they're owned by famous people? Apparently. Hood and Bloom next 'duplicated' metal spoons and cups and asked six-year-olds to rate whether they liked the original or the copy more. In one case, they told the children that the items were made of precious metal. Eighty-two percent valued the two items equally, so it seems that they bought that the items were materially identical. In the other case, the children were told that the original items had been owned by Queen Elizabeth II. Compared with the first case, more than three times as many children preferred the original. So somehow mere ownership by a queen lends the metal additional worth.

I asked LaGrone if she would have been content with a reproduction of Lennon's piano. 'Maybe it's just a trick that your mind plays on you, that *wow, it was really there* [with Lennon], but no, I would not have felt the same if it was a replica.'

In his book *The Man in the High Castle*, Philip K. Dick uses the term *historicity* to describe the intangible worth of collectibles. One character holds up two cigarette lighters and says historicity is:

> 'When a thing has history in it. Listen. One of those two Zippo lighters was in Franklin D. Roosevelt's pocket when he was assassinated [*sic*]. And one wasn't. One has historicity, a hell of a lot of it. As much as any object ever had. And one has nothing. Can you feel it?' He nudged her. 'You can't. You can't tell which is which.'

Instead of *historicity*, psychologists often use the term *authenticity* to describe the invisible value an object gains from its provenance. Recently, four researchers – Susan Gelman and Brandy Frazier at the University of Michigan and Bruce Hood and Alice Wilson at the University of Bristol – explored adults' evaluation of authentic objects in the lab. Subjects compared authentic and inauthentic items within four different categories of authenticity. There were original creations (for example, the very first lightbulb versus an everyday bulb), objects associated with famous people or events (Jackie Onassis's sunglasses or a chunk of the Berlin Wall versus your neighbour's sunglasses or a piece of random concrete), objects from a distant place or time (a moon rock or an ancient pot versus a stone from your garden or a lemonade jug), and objects associated with people or events of personal significance (your grandmother's engagement ring or your graduation robe versus an unworn ring or robe). People rated the objects on scales designating how much they'd want to touch them, own them, or place them in a museum, and how much they're worth in cash. (When people wrote *priceless*, the experimenters replaced the answer with a very high number – infinity tends to mess up stats calculations.) Every type of authentic item was rated higher than the inauthentic ones on all scales. Well, except for the personal items on the museum-worthiness scale; it takes a special type of narcissism to expect the Victoria & Albert Museum to treasure your lucky undies.

Gelman and Frazier have also explored children's appreciation of authentic objects. At age four, children decide that items with famous associations such as Ernie's rubber ducky in *Sesame Street* or the US president's favourite flag pin belong in a museum more than do their sister's ducky or their dad's pin. And at five they think original creations such as the first bicycle belong on display more than brand new objects do.

Gelman told me about taking one of her children to a rock concert: 'After it was all over and the performer had gone home, my son ran over to the stage, touched it, and ran back with a big smile on his face. Touching the stage that the performer had been on seemed somehow satisfying.' LaGrone, who doesn't consider herself especially superstitious, said of the piano, 'I hate to say it's like a good luck charm, but it's like petting Buddha.' She laughed. 'Rub the Buddha belly!'

Authentic objects are sometimes said to contain a special essence, and the belief that objects have hidden defining qualities is called *psychological essentialism*. The concept of essentialism goes beyond the placebo effect, which could explain a belief that, say, wearing Mr Rogers's cardigan will make you friendlier. In one study, researchers asked American adults to picture someone *else* wearing one of Mr Rogers's cardigan without knowing it had belonged to the beloved US children's TV presenter. Eighty percent of subjects said that there was at least a 10 percent chance that Mr Rogers's cardigan would make the oblivious wearer friendlier – and that it would be due to a transfer of 'essence'.

And you know magical thinking runs deep when even Richard Dawkins, a fellow sceptic and the author of *The God Delusion*, falls prey. As Bruce Hood pointed out to me, in the documentary *The Genius of Charles Darwin*, Dawkins endorses historicity by picking up a preserved pigeon at the Natural History Museum in Tring, England, and remarking, 'It's a very weird feeling. These are actually Darwin's own specimens.' (Later in that documentary, Dawkins pulls a book off a shelf and says, 'This is the most precious book in my collection. It's a genuine first edition *Origin of Species.* . . . This book made it possible no longer to feel the necessity to believe in anything supernatural.')

But psychological essentialism does take a few years to develop. Gelman e-mailed me that 'we are born with the underpinnings or 'ingredients' of essentialism (although perhaps not full-blown essentialism from the start).

'And yes', she went on, 'I do think this explains why we prefer authentic things, including autographs of famous people, original works of art, and Britney Spears's chewed gum.'

Magical Mystery Tour

Lennon bought the brown Steinway upright piano that inspired such a strong reaction from LaGrone in December 1970 and installed it in his home in Berkshire, where he and Yoko Ono wrote and recorded music. The next year, he released 'Imagine' on his album of the same name.

The song has since taken on a life of its own. Amnesty International uses it in promotional videos. It's sometimes played in Times Square before midnight on New Year's Eve, where thousands of people sing along. Ben & Jerry's sells the ice-cream flavor Imagine Whirled Peace. In a 2006 interview, the former US president Jimmy Carter said, '[I]n many countries around the world – my wife and I have visited about 125 countries – you hear John Lennon's song "Imagine" used almost equally with national anthems.'

George Michael purchased the 'Imagine' piano at auction in 2000 for £1.45 million. Caroline True is a music video producer and friend of Michael. 'I've been around that piano for a long time, and there's always been some little magic thing about it', she says. 'It's got John Lennon's cigarette burns on it, and I've always seen it as this thing that, "Oh my God, John Lennon wrote this song on." People used to come into his office and say, "Wow, that's amazing. I just want to touch it or have a bit of it." There was just something spiritual and peaceful and kind.' So in 2006, Michael decided to send the piano on tour, visiting those who might benefit from a message of hope.

In addition to Oklahoma City and Virginia Tech, the piano made appearances at the locations where Abraham Lincoln, John F. Kennedy, and Martin Luther King Jr. were shot. 'I never went anywhere saying this is a magic piano and it's going to cure your ills or your ails', True says, but after playing it, people often came away shaking and crying. 'There were a lot of people who said they were surprised it affected them as much as it did.'

The last stop on the tour was New Orleans. 'Things were still pretty dark and dismal then', LaGrone says. Neighbourhoods still lacked power and potable water. Those who'd spent several miserable days trapped on their roofs or in the city's massive sports arena when the levees broke still had raw feelings. True arrived without a specific game plan and came upon the museum. A local musician and huge Beatles fan named Kenny Bill Stinson was set to perform at the museum the next night, so LaGrone made Lennon's piano part of the show and got word out. About five hundred people came. 'It gave them so much joy, and it was so uplifting for them to see something that was such an important icon to them really bring along a new day', LaGrone says. '"Imagine" means so much to people, and it was almost like you were there with Lennon. It was pretty heavy.'

Why in the world would it seem that being with a piano was like being with a person who once played it decades ago? More generally, why do we value objects for their histories and assume invisible properties? What's behind essentialism? The answer may lie in disgust.

The Ick Factor

In 1890, the Scottish anthropologist Sir James George Frazer published the first edition of his extensive study of mythology and religion around the world. In *The Golden Bough*, he developed the concept of 'sympathetic magic' – a principle by which 'things act on each other at a distance through a secret sympathy' – and further divided sympathetic magic into two branches: homeopathic magic (the law of similarity) and contagious magic (the law of contact or contagion). According to the first, like produces like; cause resembles effect. That's why voodoo dolls are human shaped. (We'll pick that law up in the next chapter.) According to the second law, once in contact, always in contact. In some cultures people use hair or nail clippings as tools for affecting the person who grew them, assuming that the body parts remain in magical sympathy with their owner. And Frazer reported that even in England some people would attempt to keep wounds from festering by methodically cleaning the offending weapon *after* the injury.

In the 1980s, the psychologists Paul Rozin and Carol Nemeroff and several collaborators began churning out data suggesting that belief in the laws of sympathetic magic was still alive and well among the most modern cultures. In one study, they asked Americans to rate various experiences from –100 ('the most unpleasant thing you can imagine') to 100 ('the most pleasant thing you can imagine'). How would you feel wearing a brand new jumper? What if it's been worn by someone else and then washed? How would you feel biting into a fresh burger or an apple? What if you're biting into an existing bite mark? How would you feel using a new hairbrush? What if it was owned by someone else (and cleaned)? For each of these four objects (jumper, burger, apple, brush), the 'someone else' took four different forms, and people answered once for each scenario. What if the previous user was a close friend? A lover? A disliked peer? Or 'an unsavoury character'? Across all four objects, contact with a friend or lover rendered the experience neutral, but contact with a disliked or unsavoury character put it between –55 and –95. If a hairbrush has been disinfected, it really shouldn't matter whether you happen to get along with the person who used it before you. Yet there's still something icky about using it when you know a sworn enemy had it first.

Rozin and Nemeroff began to draw connections between the way we think about the transmission of essences (what is sometimes called *cooties* in American usage) and the way we think about biological contamination. In one study, people said they'd be none too happy to put on a jumper after it had been worn (and washed) by someone infected with tuberculosis, someone who'd lost a leg in a car accident, or a convicted murderer, and these three aversions were correlated; that is, if you feel strongly about one, you probably feel strongly about the others. According to the law of contagion, the qualities that are transmitted through contact can be physical, psychological, or moral, and can be either positive or negative. In this case, disease, misfortune, and evil apparently all obey the same laws of transmission.

The commingling between our ideas about high-and-mighty essences and down-and-dirty germs suggests a common origin, and one can guess which evolved out of which. (I'll give you a hint: would our caveman ancestors have survived if they worshipped the very first wheel because of its historicity but didn't shy from a phlegm-covered pelt?) Negative contagion elicits reactions of disgust, and although we might say an amoral person *disgusts* us, disgust originally served a very specific health function.

Charles Darwin wrote that disgust, 'in its simplest sense, means something offensive to the taste', and in the 1940s the Hungarian psychologist Andras Angyal defined it as an emotion defending the mouth from human and animal waste products. *Dis-gust* literally means 'bad taste'; the universal human expression of disgust involves closing the nostrils and opening the mouth, which would facilitate the ejection of food; and the most common physiological response to disgust is nausea, the first step to vomiting. Disgust is for keeping bad things that can make us sick out of our bodies.

But that's just disgust's origin. Rozin, Jonathan Haidt, and others argue that this physiological avoidance (and voidance) mechanism has been co-opted for other uses. It's a powerful emotional trigger that abruptly and efficiently tells us to stay away, and there's no end of threats both physical and abstract worth learning quickly to steer clear of. In line with this theory, research shows a guttural grounding to our most lofty standards of ethical behaviour. In one study, watching a disgusting film clip caused people to judge violations of moral purity – such as being promiscuous or buying sexually explicit music – as especially wrong and to judge virtues – such as meditation and abstinence from drugs – as especially righteous.

How does the conflation of spiritual and biological purity lead to belief in sacred objects? We have this emotional trigger for avoiding harmful substances such as excrement, decay, and bacteria, and we have a mental system for tracking their transmission – we pay close attention to where our food came from, who touched it, and how fresh it is. So if avoidance of moral harms buddies up on the bad-taste disgust trigger (possibly through a *bad is gross* metaphor), it's not outrageous to expect abstract taint to follow the same laws of corruption-through-contact as biological taint. And it seems that it could: both germs and psychological essences are invisible, purportedly have serious and mysterious effects, and in many cases can be identified only by following the history of their vessels – say, a piece of food or a jumper. Provenance is prioritized. Now, although we're naturally hyper-attuned to danger – rancid meat harms us more than fresh meat helps us – there's no reason positive essences wouldn't follow the same laws as negative ones. In support of this disease-disgust model of essences, researchers found that the more sensitive people are to *physical* contagion, the more they would want to purchase a jumper worn by their favourite celebrity, demonstrating sensitivity to *magical* contagion.

Evolutionary psychology is a tricky business. Critics often call it a collection of just-so stories. Granted, if you want to argue that a particular behavioural tendency exists because it aided our ancestors in survival and reproduction in such and such a way, you can't rely on the same type of evidence that most experimental psychologists use. You can't run human evolution over again in the lab a hundred times while tweaking certain variables. So while there's strong evidence to suggest

magical contagion grew out of our sensitivity to biological contamination, one can offer alternative accounts.

The easiest explanation is that it's direct mental association. The
piano simply *reminds* us of Lennon. But, Rozin and Nemeroff wrote in
one paper, 'why would many individuals become more upset by wearing an innocent-looking sweater that was once worn by Adolf Hitler
than by holding a book written about him, with his name and picture
all over the cover and the story of his life inside? The explanation of
"stronger association" begs the question of why brief contact should
result in such strong associations.'

Stronger associations from contact could result from our use of
contact to express intimacy. 'It's why we feel the need to touch the pop
star, to shake the hand of the politician', Hood says. 'Physical contact
presumably has all the legacy of biological contamination but also
reflects primate social behaviour, which is to groom and to hold and to
hug and all that sort of thing.'

'I have a scarf from a panel I did with the Dalai Lama', says Dacher
Keltner, a psychologist at the University of California, Berkeley. 'And we
have our children's hair or whatever, and those are triggers of memories
of real experience. Touch communicates trust or benevolence or cooperative intention, so any kind of object that is aligned with moral virtue,
by putting it on, the tactile contact communicates that prosociality.'

To touch someone of significance also elevates your status. 'If I
get my favourite rock and roller's drumstick', Keltner says, 'I think,
"Well, I'm with this supercool guy now, and I get all the benefits that
he enjoys." That isn't a contagion argument; it's more of an association
argument. "Hey, this is my network."'

Let's extrapolate and say that if touching John Bonham's drumstick makes you all, 'Check out who I roll with', touching exotic or historical artefacts makes you all, 'Look at what I have access to. I'm a
VIP.' Or maybe it makes you feel like you were there: 'I have such an
intimate connection with these artefacts, I practically went to the moon
and invented the lightbulb and broke down the Berlin Wall myself.'

Alternatively, belief in magical contagion extends from an awareness that properties of all kinds (heat, filth) travel through contact. Whatever the roots of magical connections, symbolism can tweak them. 'I can't imagine Einstein's fridge being a particularly important artefact', Hood says. Sure, you may touch a fridge he used frequently and feel closer to him, but wouldn't you feel closer to his essence, and perhaps even a bit cleverer, using one of his pens? Maybe not if he were a chef, but what makes him special to us – his theories and equations – was expressed through his writing implements, and so those seem to embody him especially well. Would you pay more for the chalk he first wrote '$E=mc^2$' with, even if he held it only once, or a toilet he sat on daily? I thought so.

Bless This Meal

The most profound way to incorporate something's essence is to ingest it. Hence the phrase *You are what you eat.* The Hua of Papua New Guinea believe that eating fast-growing plants will make you grow faster. And various tribes around the world have eaten the hearts of their enemies to obtain the foes' courage or strength. To test contemporary Americans' understanding of the adage, Nemeroff and Rozin told university students about a fictional tribe that hunts wild boars and marine turtles. Half the students were told that the hunters killed the turtles for their shells and the boars for food, and the other half were told that the hunters killed the turtles for food and the boars for their tusks. People tended to believe that the boar-eaters were more hairy and irritable than the turtle-eaters.

Magical thinking might even encourage vegetarianism. A study found that compared to omnivores, vegetarians more strongly endorsed claims that eating meat increases aggression, dulls thinking, and 'arouses more animal instincts in people'. (The researchers suggest, however, that perhaps belief in essences per se does not change eating habits; maybe the same need for order and meaning that leads to magical thinking also leads to vegetarianism as an ideology.)

Everyday reasoning about becoming what you eat can be taken to grotesque extremes. Armin Meiwes had a fairly normal childhood in Germany, until his father left him. He felt isolated and began fantasizing about consuming his friends as a way to become better connected. In 2001, at the age of thirty-nine, he found a willing participant online, a man who dreamed of being eaten alive. Presumably, shopping Craigslist personals for groceries was an innovation the designers of the Internet had not anticipated. The men got together, the one ate the other, and Meiwes is now locked away, as is the graphic video the two filmed of the encounter. 'He had to sacrifice his life so he could carry on living in me', Meiwes told the TV news programme *60 Minutes*. 'I got the feeling that I was actually achieving this perfect inner connection through his flesh.'

Many people have felt similar intimate connections through organ transplantation. Even if they don't know anything about the donor, they sense changes in themselves and assume they've taken on attributes of the organ's previous owner. In 1988, Claire Sylvia was a forty-seven-year-old woman dying of primary pulmonary hypertension when she received the heart and lungs of an eighteen-year-old man who'd died in a motorcycle accident. According to her book, *A Change of Heart*, in the hospital a reporter asked her, 'Now that you've had this miracle, what do you want more than anything else?' and she was surprised to hear herself reply, 'Actually, I'm dying for a beer right now.' She didn't even like beer. Sylvia discovered other new cravings: Snickers bars, green peppers, fried chicken. And her personality became more masculine, assertive. Her daughter asked her why she was walking like a burly athlete. Sylvia began to suspect that she'd assumed part of her donor's personality. She had a dream where she and a man kissed: 'I seemed to inhale him into me in the deepest breath I had ever taken. I felt like [he] and I would be together forever. When the dream was over, something had changed. I woke up knowing that [he] was my donor and that some parts of his spirit and personality were now within me.'

Sylvia is not alone. According to one Israeli study of thirty-five heart recipients, 34 percent thought they might have picked up new character traits with their new heart. Four of them credited their increased sex drive to the donor's mojo, with comments such as, 'Apparently, the donor must have been quite a guy. He must have had several women.' In a survey of organ transplant patients at UCLA, 24 percent stated explicitly that receiving tissue from a nonhuman would change their appearance, personality, or eating or sexual habits. Many organ recipients declare a preference for a donor of the same sex. Men are particularly fearful of becoming more effeminate. While some people attempt to give scientific explanations for their concerns by citing a transfer of genes, others describe a transfer of essence.

One of the interesting things about the 'once in contact, always in contact' rule is that influence can also travel from an object back to its source. In the study I mentioned earlier with the hairbrushes and the unsavoury characters, people also imagined giving a lock of their hair or one of their thoroughly cleaned hairbrushes to different people. Knowing a friend or lover has one of these items scored about a 20 on the unpleasant-to-pleasant scale. Knowing a bad guy has it scored about −15. Apparently, people are concerned about where items still 'connected' with them end up. Could it be used against you? If backwards contagion is not magical, I don't know what is. (Of course, being scientists, the authors temper their conclusions: 'Our findings do not actually indicate a belief in backwards causation; they simply report effects consistent with it under suitable conditions.' In other words: maybe there's a better explanation for these results, but we can't think of one, so good luck trying.) Nemeroff told me, 'If I use a hairbrush and then I discard it, the idea that you can somehow act on me through that hairbrush is really simple because the hairbrush and I are still connected. This rests on the assumption that there is no separation of space and time. The hairbrush and I were in contact – we merged, we shared essence, my essence is in the hairbrush. At that mystical level where all is one, acting on it is acting on me.'

In the Israeli heart transplant study, people were asked to describe their most- and least-preferred imagined donor. Half the participants made decisions based on racial, ethnic, or gender prejudices. Given such concerns about forwards contagion (incorporating traits of a source), I suspect that the phenomenon of organs donated with racist strings attached – some people specify their organs for white users only – can be explained partially by fears of backwards contagion: a bigoted donor wouldn't want a black kidney recipient to have any kind of sway over him.

Cannibalism isn't the only way to eat someone. There's also *consubstantiation*. Theologically, consubstantiation, like transubstantiation, is a process by which bread and wine become the body and blood of Jesus. In their research, Rozin and colleagues have used the term to mean simply consuming food that someone else has touched. Having someone touch your food is not the same as having someone become your food, unless you plug the law of contagion into the equation. Belief in consubstantiation plays a significant role in directing the food-sharing practices in many cultures. In India, people will readily accept food from someone of a higher caste, but eating food given by someone of a lower caste arouses fears of moral contamination and a loss of status. No one wants to incorporate the essence of a 'lesser' person.

The preparation of food is an even more intense form of handling food. You often hear that food tastes better when Mum makes it. Or a meal cooked by a partner or family member has been made with love. Someone else put blood, sweat, and tears – metaphorically if not literally – into your food, and you're enjoying the fruits of their labours. The intent has been baked right in.

From Soil to Sale

The law of contagion may play a role in geopolitical disputes as well. Research by Rozin and Sharon Wolf suggests that many consider the soil Israel sits on to be literally steeped in history, which might hinder a Jewish-Arab agreement. Among both Israelis and Jewish Americans, the researchers found a correlation between belief in magical contagion – judged by how strongly people felt inanimate objects such as a great-grandmother's ring could contain spiritual essences – and attachment to the land. The Mount Herzl cemetery, home to the bodies of several former presidents, prime ministers, and Jewish leaders, is particularly precious – and resilient to sacrilege: a quarter of Israelis said they would not give up the land under any circumstance, even if all the graves were excavated and a Palestinian prison sat there for a decade. Rozin, who has some Israeli dirt on a shelf in his office, notes similar heated territorial disputes in Northern Ireland, Iraq, Sri Lanka, and Taiwan.

Several companies have made efforts to market soil bearing essences. For example, you can buy sixteen ounces of Israeli soil for $40 (about £25) from HolyLandEarth.com. Suggested uses for the dirt include burials, groundbreakings, good-luck charms, and plantings ('*Holy Land Earth* brings the essence of the Holy Land to all living things'). Say you're not the religious type. The Auld Sod Export Company (OfficialIrishDirt.com) offers Irish soil for similar occasions: 'own a little piece of Ireland no matter how far from the Emerald Isle you are'. By their reckoning, you can take the soil out of Ireland, but you can't take Ireland out of the soil. And LittleAfrica.com sells a decorative container with sand from a beach in West Africa, 'where over 10 million of our ancestors last set foot on the African continent before boarding ships to the Americas as slaves'.

For Yankees fans, the old Yankee Stadium was hallowed ground, and the fire sale that took place after the last baseball game there in 2008 epitomizes the monetization of sports memorabilia. The history of the place seeped into every last element of the building. Sure, there's an obvious (if magical) appeal to owning a first base or a stadium seat. But dealers wanted to wring every last bit of money from the material, so fans and collectors also had the privilege of paying $3000 (about £2000) for genuine, certified, used Yankee Stadium urinals. Porcelain baptized by fire water.

Similarly, before the demolition of Chesterfield Football Club's old home, Saltergate, even the players' foot bath sold at auction (price tag: £25).

Contagion also affects commerce in more mundane ways. For example, people are more likely to purchase a T-shirt if it's been tried on by an attractive stranger of the opposite sex. (They're also less likely to buy a shirt if it's been touched by an average-looking stranger.) So if you're a manager with Abercrombie & Fitch, which is renowned for its beautiful employees, make sure your staff get tactile with the merchandise when they're working the shop floor.

Good Germs

Have you ever shared a toothbrush with a lover? Some people do it on occasion. After all, you put your tongues in each other's mouths. So you're not too worried about germs, and the gross-out factor is minimal. What about with a complete stranger? Okay, pretty gross. Now, with someone you hated? Nasty.

Carol Nemeroff has paid special attention in her research to the way beliefs about contagion influence health decisions. In one study, she asked people to draw pictures of germs of a significant other and of someone they disliked. The lovers' germs came out much less threatening-looking. You'd almost want to cuddle with them. Then she asked people to imaging spending an hour in a room with someone with the 'flu – a lover, a stranger, or an enemy – and rate (1) how likely they were to get ill and (2) if they got ill, how ill they would get. People thought the enemy would make them more ill than would the lover, with the stranger in between. We treat psychological essences as though they behave like germs, and germs as if they embody psychological essences. Yet when answering the first question – about likelihood, not intensity of illness – it made no difference who the source was. Apparently, calculating likelihood encourages a more logical processing style, enabling us to realize that the 'flu is the 'flu. But picturing *how* ill we'd get (question 2) puts us in a more emotional state, where magical thinking can creep in and influence our judgement.

If a lover's germs look cuddly, might a Hollywood starlet's germs look sexy? In December 2008 Scarlett Johansson appeared on *The Tonight Show* and reported to its host Jay Leno that after rubbing elbows with Samuel L. Jackson she had caught his cold. 'I almost don't want to lose it because it's a celebrity cold, and I feel it's valuable in some way', she said. 'How do you mean valuable?' Leno asked. 'I feel like I could sell it on eBay', she replied. On cue, Leno whips out a tissue for Johansson to blow her nose into and a bag to seal it in. The ensuing snot heard round the world earned Johansson's chosen charity, USA Harvest, $5300 (about £3400). As if to tailor the incident to my purposes, Leno's band played a twinkle sound during the bit, prompting Johansson to remark on the 'magical sound'. ('I wish it sounded like that every time', she added.)

Nemeroff notes that this form of magical thinking – allowing the personal feelings we have about a germ's source to taint our expectations of its biological potency – might make people overestimate the danger of contact with certain ostracized social groups (the homeless, criminals, gays) or underestimate the virulence of germs from a friend or lover. She's found, for instance, that when judging the risk of being infected with an STD by a partner, we tend to blur the distinction between physical risk factors, such as whether the person has been tested, and how *emotionally* safe we feel with the person. Apparently, if someone can't give you 'cooties', he can't give you crabs, either.

People also conflate emotional and physical contamination in experiences of sexual violation. In a study by researchers at the University of British Columbia of fifty women who'd had unwanted intimate contact, 70 percent reported wanting to wash themselves afterwards. That's not so strange, but women who wanted to wash had higher scores on a 'mental pollution' scale than the other women – they felt 'dirty on the inside' and couldn't get clean. A quarter of them washed excessively for several months or longer. Four of them scrubbed themselves for at least a year. These women fruitlessly employed a physical cleaning routine to expunge a moral violation. And typically, changes in the way a victim thinks about the perpetrator – through, say, forgiveness or further offences – result in changes in the severity of pollution symptoms connected to that original assault. Once in contact, always in contact.

We can also feel tainted by our own dirty deeds. For example, there's the Lady Macbeth effect: after recalling an unethical act, people feel the urge to wash their hands.

In addition to treating moral impurity as a soiling substance, we also treat luck (both good and bad) as a residue susceptible to scrubbing. In a study, gamblers who washed their hands after a losing streak increased the size of their bets, and gamblers who washed their hands after a winning streak decreased their betting.

And Matthew Desmond, the author of *On the Fireline*, told me that wildland firefighters don't wash their shirts all season. 'By the end of the season, they are blackened with soot and ash and are salty from the sweat', he said. 'Part of the reason is practical: they'd look like rookies out on the line if they showed up with a bright yellow shirt. But part is somewhat superstitious, as if washing one's shirt would wash away all the experience one amassed chasing smoke that year.'

A Coating of Pixie Dust

'As time, space, and the ether soak through all things, so (we feel) do abstract and essential goodness, beauty, strength, significance, justice, soak through all things good, strong, significant, and just', William James wrote in *The Varieties of Religious Experience*.

We've seen how belief in essences affects our treatment of germs, clothing, soil, pigeons, spoons, hearts, urinals, rings, and musical instruments. In thinking about how anything can have psychological significance, how we could potentially essentialize everything in our environment, I'm reminded of an old sketch on *Saturday Night Live*. It was a fake ad for Faecal-Vision glasses. Wearing the specs reveals all the invisible faecal coliform bacteria in the world you usually miss. In the ad, the camera looks through the glasses at a restaurant, and you see everything smeared with this glowing substance – the walls, the food, people's faces. Looking at the world through a magical lens reveals everything potentially glowing with significance, if only you considered where stuff had been. That stapler, say, was touched by the jerk down the hall, this mouse pad is from a job you hated, your trousers sat in the same seat as a perv on the train. It's creepy. So how do we deal with it all? The same way we deal with typical disgust – you just don't think about it. You just don't put on the Faecal-Vision glasses. If you stopped to consider what's on every doorknob in the world, you'd never leave the house. You might not even be happy inside your home. A former housemate didn't wash his hands after peeing, so I avoided touching the inner bathroom door handle – I washed, then opened the door, then rinsed again – but I restricted myself from going too far. I'm sure every touchable surface in the kitchen was affected by my housemate's habit, but I put it out of my mind so that I wouldn't starve to death.

Sometimes people or cultures have more explicit rituals to manage their anxieties. For example, food can never be kept completely pure, so according to the laws of kashrut, a food is kosher as long as no more than a sixtieth of its volume is composed of a tainted substance. Meanwhile, others abide by the five-second rule.

Just as what counts as contact can be circumscribed, it can also bleed the boundaries. In some societies, stepping through someone's shadow counts as physical contact, and in our own, we often joke about scrubbing our eyes after taking in a grotesque sight.

⁞⁞⁞⁞⁞⁞⁞⁞⁞⁞⁞

There is a disorder that approximates the loss of a sense of essences – sort of the opposite of Faecal-Vision glasses. Typically, person and object recognition relies upon two distinct pathways in the brain, one responsible for overt recognition and one that adds an emotional sense of familiarity. If the second route doesn't function properly because of psychiatric illness or brain injury, an individual will recognize people he knows but won't feel the warm glow about them he expects to feel. In some cases he'll interpret this lack of emotional response to mean that people close to him have been replaced by impostors or robots. This confusion is called the Capgras delusion. Capgras sufferers have also been known to suspect that familiar buildings and household items have been swapped out for near-identical duplicates. Bruce Hood wrote in his book *SuperSense* that the Capgras delusion is the result of losing one's sense that there are supernatural essences inside people, pets, and objects.

Personally, I find it fascinating that people should trust their perceptions of invisible essences more readily than their sensory organs as indicators of reality – as though the magical realm (or the emotional realm, anyway) were more essential than the concrete one. But such a prioritizing of input fits with the reports of people who enter spiritual-mystical states, whether through drugs, meditation, or near-death experiences. They interpret these encounters as brushes with an even more ultimate reality, leaving them with the feeling that it's the everyday world of the mundane that's the illusion. No one is quite sure how the brain decides what is 'real', but it seems clear that mystical experiences answer our search for the Truth in an oddly satisfying way.

Another disorder, the Cotard delusion, is hypothesized to be a more general and extreme form of Capgras, in which a wider array of sensory input fails to stimulate the amygdala and generate emotion. The broad distortion of reality leaves sufferers with only one possible answer. Instead of blaming others for trying to dupe them by swapping out particular people or objects, they see themselves as the problem. They conclude that they are, in fact, dead.

Alas, life without a sense of the sacred would be no life at all.

The Stuff of Magic

People treat essences as a substance, a force, an energy, or all three. It's the stuff of magic, and it surrounds us. In Polynesia they call it *mana* (not to be confused with *manna*). In China they call it *chi* or *qi*. In the US and the UK, we call it *vibes*. The concept is slightly different in each culture, but there's a universal notion of an invisible quality that carries luck or power or spiritual essence. It can communicate information instantly or become embedded in objects. The concept accompanies much of magical thinking. Marjaana Lindeman and collaborators found that the belief that energy can have biological or mental attributes – that it's capable of, for instance, living, dying, purifying, poisoning, desiring, believing, or being masculine or feminine – is significantly correlated with religious, spiritual, and other paranormal beliefs, as well as belief in the purpose of events.

We may not have much patience for sorcerers in the West these days, but we still open our wallets for wielders of the woo. When I first saw a flier for a feng shui workshop in the 1990s I thought it was a prank – there's an ancient mystical art of furniture arrangement? But this Chinese practice of ordering one's environment to enable the flow of positive energy has gained ground in the US. In 2004 a California state assemblyman introduced a resolution that would encourage building departments to abide by feng shui principles. And it's not hard to find a self-proclaimed feng shui 'master' who will charge thousands of dollars to tell you how to build or decorate your home.

Alternative medicine also has many adherents. Homeopathic remedies rely on the notion that certain medicines have a vital force that can remain in a solution even after all the original medicine has been diluted out. Energy healing seeks to manipulate vital energy through touch, prayer, needles, or devices such as magnets or crystals. These placebos by themselves usually cause little harm to one's health, but sometimes people use them instead of, rather than complementary to, evidence-based treatments.

And millions of people rely on astrologers in the hopes of aligning themselves with the cosmic currents. At least those not put off by the recent rejiggering of the zodiac calendar. (Speaking of which, I'm sorry, but gaining the Ophiuchus constellation as a sign does not make up for losing Pluto as a planet. Too little, too late, astronomy.)

In everyday language, we often speak of being on the same 'wavelength' as someone, or going to an event with 'bad energy'. I submit that these metaphors, used by even the most materialist of us, reveal something about how we view the actual world. Internal sources of discomfort can be mysterious, and when our Spidey senses perk up, we look to an external influence.

Now, whether or not you call it *vibes*, when something in a person, a place, or a situation makes your hair stand on end, you probably don't want to ignore it.

Embracing Enchantment

I slept with a Beanie Baby for eight years. Those years were from age eighteen to twenty-six. Maybe if I put that in context, it won't sound so pathetic.

I'll start with how I lost him. I was working at a physics lab near Chicago after graduate school. To save money I lived on campus. One benefit: housekeeping. Each week someone came to change my sheets. Typically, I removed the plush little guy from my bed before driving to the office on those days, but one day I forgot, and when I returned he wasn't there. I can only assume he got bundled up in my linens and carried off to the laundry processing plant. He was never seen again (by me).

In the midst of my ensuing exchange with the lab's housing office, I sent the following e-mail:

> You may be wondering why someone my age cares about a red
> plush dragon. When I graduated from high school I went to

Alaska by myself and lived outside a national park for three months. When I arrived in the state I bought this Beanie Baby to keep me company, and I named him Blip. I took him on all my hikes and to the tops of mountains. I took pictures of him mingling with wombats and arctic ground squirrels. I have kept him with me ever since. That was a difficult time in my life, and he reminds me of a part of myself that I discovered in Alaska. He's sort of an icon, and he brings me peace.

What else can we do to bring back Blip?

So yeah, I made a friend in Alaska.

(Here's an irony that's eluded me until now: I went up there for a job as a hotel housekeeper. They say cleanliness is next to godliness. When it comes to Blip, that's certainly true: housekeeping bringeth and housekeeping taketh away.)

Of course, I'm telling myself, *He's just cloth and beans, right? I could go on eBay and order a new one. All the same memory cues as the last one but with less nap.* Nope, that just wouldn't cut it. This particular beanbag embodied some important lessons I learned that summer, from the everyday – I'd been hospitalized for depression half a year earlier and was still learning to get out of bed in the morning – to the thrilling (but no less obvious) – don't linger atop an Alaskan peak with a snowstorm coming on while high on hallucinogens. Blip saw me through times both life affirming and death defying. He had absorbed my history and was now a part of me, and I was a part of him. A personal narrative suffused into red felt.

Treating parts of the physical world as blessed or cursed can have negative health consequences, and it can drive you mad with fear of psychological contamination, but this law of magic can be put to good use.

A friend wears boar's tusks around his neck because, he says, boars are great at focusing on what's in front of them, and the tusks keep him present – not because they act as a mnemonic but because of their 'vibrations'. Maybe that's a bit eccentric, but a recent study found

that subjects made 38 percent more golf putts when told the putter
they were using belonged to the American pro golfer Ben Curtis than
when told nothing about the club. When using objects with special
associations, credence in contagion can increase confidence and enable
one to persist at difficult tasks.

Paul Rozin simultaneously deconstructs and rationalizes magical
thinking. For example, he owns several of his mother's paintings and
lots of old books, including some that belonged to the late pioneering
social psychologist Solomon Asch. 'It gives me a good feeling', he told
me. 'A sense of their presence.' Belief in the law of contagion makes the
physical world glow with meaning and connects us to other people and
to history. It makes the universe feel less inert, less sterile, less lonely.

For me, perhaps Blip became an emotional crutch, the way a child
can't go anywhere without his blanky. But the fact that I moved on
without him, and still retain the lessons of my intense summer in
Alaska, suggests that allowing myself a little magical thinking during
my years with him didn't handicap me. He was more training wheel
than crutch.

||||||||||||||

When Lennon's piano arrived in New Orleans, Libra LaGrone was
still rebuilding her house and sleeping in her late grandfather's sweatshirt every night. 'Sitting down at the piano and playing a couple notes
gave me the same comfort and joy as sitting on my grandfather's lap
and hugging him and hearing him tell me why the sea is salty', she said.
'It made all my fears and sadness go away.'

'I had no idea it was going to have the effect it was going to have,'
Caroline True says of the tour. 'Just doing something with an inanimate object.'

2

|||||||||||||||||||||||||||

Symbols Have Power

Spells, Ceremonies, and the Law of Similarity

Growing up in the Bronx, New York, Gino Castignoli often received Yankees tickets from his neighbour, a maître d' at the 21 Club. He sat right on the edge of the field, next to the action, but the team was in a long lull – they didn't win an American League championship between 1964 and 1976 – and Castignoli didn't appreciate the attitude of the fans. So he never became one himself.

Then, in 1975, Jim Rice and Fred Lynn caught Castignoli's eye. That year, Rice and Lynn had joined the rival Boston Red Sox. At the time no one had ever won Rookie of the Year and Most Valuable Player awards in the same season, but Lynn won them both. And Rice wasn't far behind, coming in second place for the rookie award and third for MVP. They were nicknamed the Gold Dust Twins. Castignoli followed them every week in *The Sporting News*, becoming a lone Red Sox fan in hostile territory.

Still living in the Bronx, Castignoli is now a construction worker in his forties. When the Yankees decided to build a new $1.5 billion stadium, many of his friends ended up on the job. 'I was offered work there a lot, and I always refused', he told me. He wanted no part in the enforcement of the Evil Empire. 'It's like building your own tomb over there, digging your own grave, ya know?'

Then he got an idea.

'One day I said, "I could go there and jinx them."'

Jinxes are nothing new to the Yankees–Red Sox rivalry. In 1919, Babe Ruth hit twenty-nine home runs, setting a single-season home run record. He threatened to retire if the Red Sox didn't double his salary, so Harry Frazee, the owner of the team, sold him to the Yankees. Before the sale, the Red Sox had won five of the first fifteen World Series championships; the Yankees made not even an appearance. But over the next eighty-four years, the Yankees won twenty-six championships, the Red Sox none. It was called the Curse of the Bambino.

It took another heavy hitter to end the curse. In 2003, David Ortiz was a designated hitter (DH) whose contract with the Minnesota Twins baseball team had just expired. George Steinbrenner, the owner of the Yankees, reportedly told his general manager, Brian Cashman, to sign him. They passed, and Ortiz signed with the Red Sox. The next year, Ortiz played his first full season as Boston's full-time DH. His value to the team became epically clear in the championship playoffs.

Boston faced the Yankees in the American League championship series, as they had the year before. At first, it looked like business as usual, as New York pulled ahead three games to none and, late in game four, were set to win the championship. But Boston managed to tie it up, sending the game into extra play. After playing on the brink of death for three more innings, Ortiz hit a home run in the twelfth, winning the game. Game five brought no less suspense, and Ortiz once again saved the day, winning the game in the fourteenth inning with a hit that allowed another player to score a run. Boston won the next two games, becoming the first Major League team in history to pull out of a three-game hole in a playoff or championship series. Ortiz was series MVP. More important, he got them to the World Series, where they swept the St Louis Cardinals in four games, winning their first title in eighty-six years and breaking the Curse of the Bambino.

That's why Gino Castignoli finally accepted the construction job offer and went to work his first day at the new stadium wearing a David Ortiz jersey.

|||||||||||||||

Castignoli showed up on a Saturday in August 2007. 'One of the fore-men for the general contractor, Turner, who I knew for years, said, "What are you doing here? I can't believe you're coming here to work"', Castignoli recalls. 'I told him straight out, "I come here just to bury this shirt in the concrete and jinx the team." He laughed.'

At eight o'clock Castignoli removed the shirt and placed it under some steel rebar. 'I folded it nice, laid it to rest, and told everybody, "The Yankees will be cursed for thirty years. After that I really don't care what happens because I'll probably be dead." And then the guy put the con-crete on top.' Castignoli worked the rest of the day and didn't go back.

The shirt sat there, entombed, until April 2008. Right before a series between Boston and New York, a couple of workers told the *New York Post* about the burial. On Friday, 11 April, the story broke.

The Yankees didn't believe it. 'Memo to the *Post:* You're 10 days late for April Fool's Day', their PR rep wrote in a statement. Then two more witnesses came forward with corroborating details, including the location of the shirt. Workers spent five hours jackhammering a two-foot-by-three-foot hole until they spotted the shirt two feet down. They informed higher-ups, who planned a ceremony the following day to finish the job. On Sunday, the project foreman and a Turner execu-tive pulled the shirt out of the ground in front of a swarm of reporters and held it high. The Yankees auctioned it off for $175,000 and donated the proceeds to the Jimmy Fund, a charity in partnership with the Red Sox.

Castignoli says most of his Yankee-loving friends considered the stunt harmless, but he also received hate mail, and the team didn't take too kindly to his actions. 'They tried to have me arrested, and then they threatened to sue me personally', he says. 'That went south. The Yankees went to the Bronx district attorney, and he told them to beat it.' Hank Steinbrenner, the Yankees' owner's son, told the *Post,* 'I hope his coworkers kick the shit out of him.'

I went online and read baseball discussion boards from those few days in April 2008. Some fans thought it was funny, some dismissed anyone who took jinxes seriously, and many expressed anger at the trickster, often suggesting his body would make a great substitute for concrete when refilling the hole. I'm sure, had the shirt been discovered after several losing seasons, comments would have tended towards the latter. Sometimes 'curses' take a while to gestate.

A number of Yankees fans tentatively aired their superstition online. 'If it were me, I would try to get to the bottom of it just for [peace] of mind', one wrote. 'Maybe I need a life, but this shit fucking bothers me', another typed. And, after the removal: 'I don't know why, but I for one am glad it's gone.' The legendary baseball manager Yogi Berra also straddled belief and scepticism, telling the *Post*, 'I was never superstitious, but maybe we should bury one over there at Fenway too.'

The Yankees won the World Series in 2009, their first season in the new stadium, and I asked Castignoli if they would have lost had the shirt remained. 'I would like to believe that, but when you spend half a billion dollars' – they signed contracts worth about £270 million with three new players after the 2008 season – 'I don't care what you bury under there unless it's dynamite.'

But the Yankees hadn't won the title since their cluster of wins in 1996, 1998, 1999, and 2000 and apparently felt they needed more than money to get out of their funk. From one message board: 'Spending five hours to dig up a piece of cloth with their rival's name printed on it? Seems like they're paranoid and nervous in NY. They may have won twenty-six titles, but it could be a long, long time before they win one in this millennium. Digging up the Sox shirt just reaffirms the fact that they already think they're cursed.'

Ideal Connexions

Symbolism falls into the second law of sympathetic magic outlined by Sir James Frazer, the law of similarity: *like produces like*, or *an effect resembles its cause*. 'The magician infers that he can produce any effect he desires merely by imitating it', Frazer wrote. The jersey is an imitation of the enemy and so embodies him. It's voodoo science.

Frazer was hugely influenced by the work of another English anthropologist, Sir E. B. Tylor. In the 1871 book *Primitive Culture*, Tylor wrote: 'The principal key to the understanding of Occult Science is to consider it as based on the Association of Ideas, a faculty which lies at the very foundation of human reason, but in no small degree of human unreason also.' He argued that we treat 'ideal connexions' as real ones: if two things are linked associatively in our heads, we suspect a physical, causal link. Expecting physical reality to work along the same lines of influence as mental associations is magical thinking. And in humanity's desire for control over the world, we have a long history of trying to pull these sympathetic strings of influence.

You don't need to outwardly endorse voodoo to suspect its reality. Let's say I posted a picture of your mum on a dartboard and handed you some darts. For the sake of a scientific experiment, would you throw the darts at the picture? *Okay . . .* , you'd probably say, with a quizzical look on your face. Would those darts actually harm your mother? *Of course not*, you'd probably say, with a furrowed brow. Measuring magical thinking with simple yes-no questions like these proves difficult because, in answering them, rationality takes precedence. But I could look for subtle behavioural effects of irrational instincts. Are you less accurate when throwing darts at a picture of a hero than, say, Hitler?

This experiment has been carried out. Paul Rozin, Linda Millman, and Carol Nemeroff reported in 1986 that when subjects targeted pictures of JFK or someone else they liked, their darts strayed about eleven millimetres farther from the bull's-eye than when targeting Hitler or someone they hated. The difference was small but reliable. Another team of researchers reported similar results twenty years later – this time with darts thrown at a photo of a baby. And they found that people's levels of rationality played no role in accuracy, but how strongly people tended to rely on their gut feelings in daily life did.

Further adventures in photo voodoo occurred during the introduction of photography into parts of Africa. Europeans wowed the locals with various 'magical' technologies and claimed they could manipulate people by manipulating photos of them. Many Africans refused to be photographed, unless they were holding a Bible, which purportedly acted as a protective talisman.

What explains these beliefs? Do we somehow confuse photographs for reality? Perhaps. We didn't evolve in an environment that had photographs, so we don't have natural cognitive mechanisms for differentiating photos from real life. As far as our ancient brains are concerned, if it looks like something, it is something.

The ability to make even simple distinctions between appearances and reality takes a few years to develop. Infants try to grasp pictures of objects on a page. Even after children learn to tell photographs apart from reality, they might make the mistake of saying, for instance, that a picture of ice cream feels cold. I'm still working on mastering appearance-reality distinctions. I'll breathe through only my nose around stinky things because I don't want their gross particles in my mouth (a frustrating predicament with congestion), and I find that seeing even a picture of something stinky makes me seal my lips.

If we are subconsciously fooled even by a photo, imagine the greater havoc wreaked by moving images. I find horror films terrifying and sexy films arousing, even though I know they're not real. Studies suggest that watching TV can partially fulfil our need for real friends. And as for three-dimensional imitations of reality, Rozin and colleagues moulded two pieces of fudge into the shape of a muffin and the shape of a dog turd and asked subjects to rate how much they'd like to eat each one. Subjects also rated how much they would enjoy putting a rubber sink plug or fake rubber vomit in their mouths. Unsurprisingly, the fake turd and fake vomit were not big hits.

Having the half belief (or what the philosopher Tamar Gendler calls an *alief*) that an image or decoy is literally what it represents is one thing, but how might one jump from that confusion, which is usually cleared up, to conceiving the abstract law that influence can be transmitted between the representation and its referent – the law of similarity? One possibility is that there's cognitive dissonance between the alief that an image is actually the thing it represents and the explicit belief that it's not. Perhaps the juxtaposition of an alief and a contrasting belief can lead to a new belief (or alief). In this case, hypothesizing a magical means of causal transmission would accommodate both your anxiety about destroying a photo of a loved one and your understanding that it's just a photo.

But something more than a simple confusion between appearances and reality is likely going on. Bruce Hood, Paul Bloom, and colleagues asked subjects to bring childhood sentimental items into the lab. They photographed the objects but zoomed in and defocused the lens so the photos were unrecognizable to anyone who didn't already know what the objects were. Then they asked the subjects to cut these semiabstract representations in half. Electrodermal sensors revealed that destruction of the photos aroused anxiety in the subjects – more than cutting up recognizable photos of very similar items. Subjects feared harming a blurry image with a vague connection to a beloved stuffed animal.

Similarly, a sports jersey could not be confused for an actual person. And the US flag depicts what it represents with even less verisimilitude, but many Americans still feel those stars and stripes demand legal protection from immolation. After all, 'Old Glory' stands for the deaths of American men at war, the struggles for civil rights, the moon landing that was one giant leap for U!S!A! Flag burning certainly sends a message of disrespect, but if the research on belief in the law of similarity has any bearing, the act does something more than communicate a message: it creates the sense that America is being harmed *directly*. In a study of Americans and Brazilians, 34 percent of young adults said cutting up one's national flag to use as rags for cleaning a bathroom *in private* called for punishment, and 35 percent said the act was wrong *no matter one's country's customs*. That is, it's hurtful even if there's no audience to offend. Might we see flag burnings as a form of voodoo?

(Not all accusations of such voodoo require much interpretation. Before the US played Mexico in a World Cup qualifying game in

February 2009, an electronics chain, RadioShack, ran an ill-advised ad campaign. They printed coupons in a Mexican daily newspaper that readers could redeem in shops for dolls dressed as American football players. The ad suggested that fans mutilate the dolls while watching the game in order to 'Help end the losing streak so Mexico advances.' A day later the company pulled out of the promotion.)

Belief in voodoo may depend in part on the fact that the more you think about something, the more probable it seems. In a study conducted before the 1976 presidential election, Americans who were asked to imagine Jimmy Carter winning gave him better odds than those asked to picture Gerald Ford winning. According to the Nobel Prize–winning work of Amos Tversky and Daniel Kahneman, people have a mental toolbox full of *heuristics*, or rules of thumb. The *availability heuristic* relies on the fact that if something happens a lot, examples will come to mind easily, and it reverses this association, making us believe that if something comes to mind easily, or is pictured vividly, it must happen a lot, or be more likely to happen. (Sometimes the heuristic fails us, as when we overestimate the danger of air travel because plane crashes make the news and become imprinted in our memories.) By definition, symbols (including suggestive acts) call to mind what they represent. And if they make a particular outcome more vivid and available in our minds, that outcome will seem more likely to occur. Now, because *exposure* to symbols makes things *seem* more likely, we may jump to the alief that the *existence* of symbols actually *makes* things more likely.

Alternatively, or additionally, we might apply a like-causes-like heuristic, explored below, wherein similar causes and effects seem to go together, even if the similarity is only conceptual.

Like Causes Like

In a classic paper called *Craps and Magic*, the American sociologist James Henslin reported that gamblers will often throw dice harder when they want a high number, as if the amount of force translates into the quantity of dots showing on a die. It's an example of belief in the law of similarity, like producing like.

The law of similarity has solid roots. In the 1843 text *A System of Logic, Ratiocinative and Inductive*, the philosopher John Stuart Mill notes a strong 'natural prejudice' to believe that 'the conditions of a phenomenon must, or at least probably will, resemble the phenomenon itself'. We assume that causes resemble their effects.

Mill went on to explain that our intuitions regularly match reality: 'The cause does, in very many cases, resemble its effect; like produces like' – which could explain how we pick up our prejudice. He notes wax impressions (where the tool matches the indentation), moving bodies setting others in motion (a body travelling one way will knock others in that direction), illness contagion (one sick person makes others sick), plants and animals resembling their progeny (like father like son), memories of sensations resembling the original sensations (remembering happy times makes you happy), and the spread of emotions (laugh and the world laughs with you). Think also about how red crayons draw red lines, how a cold ice cube cools a drink, and how big lightning causes big thunder.

What kind of effect would resemble injury to a photo of your mother? The first that comes to mind is injury to your actual mother. And if one wanted to bring rain to one's crops in a dry season, what kind of cause might resemble such an outcome? The classic voodoo technique, as Frazer notes, is to pour water on one's field.

We expect like to cause like in nearly every endeavour, even if cause and effect are alike only in very abstract ways.

|||||||||||||||

The law of similarity pops up in both magical and prosaic thinking. In an article titled 'Like Goes with Like: The Role of Representativeness in Erroneous and Pseudo-Scientific Beliefs', the psychologists Thomas Gilovich and Kenneth Savitsky laid out several ways the law has overstepped its bounds. In the last chapter, I discussed 'you are what you eat', with people assuming boar-eaters would become more boorish than turtle-eaters. Similarly, common misconception holds that eating greasy snacks directly leads to oily skin and acne, drinking milk produces phlegm, and spicy food is the source of heartburn. And as a child I curtailed my consumption of Muenster cheese when my parents warned that eating too much could turn one into a monster. The available evidence has not borne out this claim, either.

Gilovich and Savitsky also note the New Age 'rebirthing' movement, in which people try to counteract the circumstances of their nativity by reenacting their births. Someone born feet-first might feel he's always going in the wrong direction in life, and someone born via caesarean might feel she can never finish things for herself. In graphology, personality is inferred from handwriting. In one study, judges assumed, for example, that compact handwriting symbolizes introversion, ascending slope signals optimism, angular lines indicate an analytical personality, and a regular rhythm displays reliability. All false correlations. Astrology leans on similarity too. I was born an Aries, and the Aries constellation is a few dots in the sky with a forced resemblance to a ram, and rams symbolize stubbornness, so I'm supposed to be stubborn. (Which I often am, but I'm just as frequently easygoing.) The psychologists Richard Nisbett and Lee Ross note that we sometimes show hostility to explanations that don't fit 'like causes like'. For a time, people ridiculed the idea that something as small as a mosquito could inflict something as devastating as yellow fever.

For hundreds of years people have used plants that resemble body parts to cure those body parts. Walnuts for the brain, for example. (In the 1600s the German mystic Jakob Böhme popularized the Doctrine of Signatures, writing that God marked flora with a sign or signature indicating its intended purpose.) Even today, studies of pill colour show that people expect red, yellow, and orange medications to act as stimulants, perhaps because of their perceptual similarity to fire. Meds in blue and green – the colours of sky and grass and sea – are thought to be chill pills. Except in Italy, where blue pills invigorate men. (And not just the little blue pill that invigorates men everywhere.) Researchers suggest this culture-bound effect might have something to do with the fact that their national football team, known for its energy and prowess, is the *Azzurri* – Italian for 'blue'.

Belief in the law of similarity has the power to infiltrate every aspect of our interaction with the world. Because we think symbolically and associatively, we have the capacity and the compulsion to see everything as a representation of something else. And if we treat symbolic relationships as real, physical, ones, the universe becomes a web woven with invisible threads of influence mirroring our own associative networks. Human meaning then becomes a mediator of causality, and the whole world becomes magic.

Wiccan Wisdom

A few years ago I wrote an article about cold feet before weddings. I interviewed a bridal counsellor named Allison Moir-Smith who had her own struggles before the big day and realized that a deep sense of loss often accompanies the engagement period. You're giving up something huge and familiar – single life – and such a sacrifice requires acknowledgement and even mourning before you can move on. Among her preparatory exercises, Moir-Smith recommended private rituals, including this piece of mental sorcery: write down a list of everything you'll be leaving, ponder it, and then burn it. It's a psychological tactic, one whose effectiveness suggests but doesn't require a belief in witchcraft, and I got to wondering if the sceptic could borrow other practices from the occult sciences for his own therapeutic benefit.

I called up Steven Farmer, a shamanistic practitioner and former psychotherapist, who said 'release ceremonies' like Moir-Smith's are popular. 'I'm moving soon. Let's suppose I want to release any attachment I have to the house I'm in. I might do something like go out on the land, find just a small stone that feels right, bring it inside, and write out what it is I want to release,' he told me. 'I might write 'I, Steven, choose to fully release any and all attachments to the land and to this house', and wrap the stone in the paper and then put it under my pillow.' After sleeping on it, he would go down to the beach, burn the paper into a small hole in the sand, then scoop up the ashes and rock and throw them into the sea. Finally, he would thank 'grandmother ocean' for taking the package, and thank 'grandfather fire' for his contribution, but presumably these flourishes are optional.

Recently, a team of psychologists demonstrated the therapeutic potential of rituals with symbolic content, or what they call *metaphor therapy*. When people wrote down things that bothered them – regretted decisions, unsatisfied desires, reactions to a tragedy – and sealed them inside envelopes, they gained emotional relief. Envelope closure led to psychological closure.

||||||||||||||

Secular rituals permeate our lives, from handshakes – the corporal strength of which can signify and even solidify the intangible bond of a deal or a relationship – to Nobel Prize Award Ceremonies, whose ostentation both indicates and perpetuates our esteem for cerebral achievement.

'Ritual practices signify to the world and the person a transition from one stage of life to another, whether it's a marriage, a bar mitzvah, or a naming', says Gail McCabe, a sociologist and 'humanist officiant'. Humanism, as she defines it, is 'a naturalistic scientific secular philosophy of life that precludes any belief or reliance upon supposedly supernatural powers'. McCabe creates and executes nonreligious ceremonies with her clients to mark the menopause, divorce, birth, and marriage.

Weddings are one of the most secure vehicles for the transmission of tradition. Although McCabe doesn't mention God or Vishnu or Xenu or Thor in her services, many of her weddings end up quite orthodox – the black tux and white veil intact. And where there's tradition, there's superstition. (Which is the smoke and which is the fire?) According to one of the most common admonishments, the groom should not see the bride before the ceremony, dating back to the days of arranged marriages and the real risk of the man spotting an unseemly costar and running off solo. Other traditions have more transparent symbolic value: a bride should wear something old to connect to her family and history, something new to invite a bright future, something borrowed to feed off the good fortune of a happily betrothed friend, and something blue to represent purity. And rain on your wedding day – well, a stormy ceremony could easily portend a stormy marriage, with precipitation precipitating tears, but we've prophylactically reversed that little omen so bad weather now means good luck. (And good weather still means good luck because, hey, it's sunny and your pictures will look great.)

Getting off on the wrong foot is a constant source of concern with both real and symbolic consequences in many endeavours, from friendships to athletic competitions. That may be why we invest so much in initiation ceremonies: we want to set the emotional tone for a journey. We may also feel, through the magical law of similarity, that future turns on that journey will imitate those first few steps, so we'd better get them right.

At the university I went to, a big set of gates opens twice a year, once in the autumn for freshers to enter and once in the spring for graduates to exit. This was a liberal institution, but the gates held sacred power, and one didn't want to botch one's entrance. It was said that if you entered twice, or not at all, you wouldn't graduate. I never walked in. In my first year I ended up taking advantage of the new-found freedom and social opportunities not available at my small boarding school, and in the process I missed class meetings and assignments, thus ending my first term with three poor marks and one incomplete. Not a good way to start university. The logical explanation is that whatever made me miss classes also made me miss my opportunity to walk through the gates (I probably slept through it), but the thought at least occurred to me that year that flubbing the ritual entrance destined me to underachievement.

||||||||||||||||

My mother never removed her wedding ring for the entire thirty-eight years of my parents' marriage. I asked her one time if that continuity symbolized anything. She said it stood for the continuity of their relationship. So would removing it break the bond? *No, of course not*, she said. Well, then, how would she feel if it came off? *Anxious*. Why? She didn't know.

We may see symbolic connections between objects or acts and larger ideas such as commitment because, according to some scientists, metaphors form the natural language of thought. We get a handle on the abstract via the real and the specific. According to the principles of *embodied cognition*, all of human thought finds grounding in our physical experience in the world, and metaphors are not just linguistic conventions – they go deeper. Such expressions as 'the prices rose' (more=up) and 'she's a warm person' (affection=warmth) may originate in the corporeal experiences of building piles and hugging, respectively. 'The brain computes a circuit between more and up and affection and warmth and all kinds of other cases', says George Lakoff, a linguist at the University of California at Berkeley. 'You learn what are called *primary metaphors*, without using language, just by functioning in the world.' Important is big. Understanding is grasping. Desire is hunger. He estimates that by the time you were five, without being aware, you probably learned hundreds of primary metaphors.

Exploring the link between affection and warmth, one study found that after holding hot coffee we judge strangers friendlier. Another found that people take job candidates more seriously when rating them while holding heavy versus light clipboards. (Important matters are weighty.) And you know how tasks can be either rough going or smooth sailing? After playing with rough-textured puzzle pieces, people evaluate ambiguous interactions with others as more adversarial than after fingering smooth pieces.

Primary metaphors fit together into complex metaphors like atoms in a molecule. For instance, *a relationship is a shared journey* contains *intimacy is closeness*, *purposes are destinations*, and *difficulties are obstacles*.

If we base abstract meaning on concrete experience, perhaps that's why we tend to infuse concrete experience with abstract meaning. Certain events act as metaphors for other, often larger, phenomena, and we then mistake the metaphors in our heads for real-world associations. It's like causing like, where cause and effect are alike only by virtue of abstract analogy. So, by not properly entering my university campus through the front gate, I feared I had not properly entered the overall university experience.

The Name Game

Mental associations tend to run bidirectionally: lightning brings to mind thunder, and thunder brings to mind lightning. And because the law of similarity dictates that external reality obey our internal associations, we expect physical causality to work both ways too. An icon can transmit your actions to the real McCoy (typical voodoo doll stuff), and meanwhile the essence of the real McCoy can be transmitted into the icon – as well as whatever it touches. That's why people feared the Ortiz jersey buried in the stadium. They expected the power of Ortiz and the Red Sox to work its way onto the field through the shirt.

Another example of drawing essences from imitations comes from the music world. In the last chapter we saw how people felt empowered by touching a piano a legend had played. Well, many guitar players fetishize instruments that merely *look* like the ones their rock idols strummed. Apparently, they believe musical mojo goes into the original axe through the law of contagion, then is broadcast to any guitar that resembles it through the law of similarity – and can be drawn out by any fan. Dickey Betts of the Allman Brothers Band told a researcher that his 1954 Stratocaster, identical to Buddy Holly's, allows him to 'channel' Holly because it has the 'soul' of the original.

Many other fetish objects, including holy icons, hundred-dollar bills, and high heels, are worshipped not for their functional utility but because of what they represent (in the examples cited: gods, goods, and gams).

In both Africa and Europe, people have placed devotional images in water and then drunk the water for its healing power. Islamic scholars have written Qur'anic verses in chalk, then washed the chalk off the board and given the water to ailing patients. And during communion, Christians take bread and wine they believe to embody, through the law of similarity, the flesh and blood of Christ.

Advertising may owe something to magical thinking too. Logos and brand names act as icons for ephemeral qualities such as luxury or edginess or power. Do we believe a tiny bit that, say, the charging bull on the Red Bull can somehow energizes the fluid inside? And maybe the swooches on your trainers make you feel you can really zoom.

The influential Swiss psychologist Jean Piaget argued that young children regularly treat names – one form of representation – as part of reality, a phenomenon called *nominal realism*. In Rozin, Millman, and Nemeroff's 1986 paper (the one with the dartboard and the fake vomit), they reported evidence for nominal realism in adults. An experimenter placed two empty bottles in front of each subject and filled them about a third of the way from a box of Domino Sugar. Then the experimenter handed the subject two peel-off labels, one that said SUCROSE (TABLE SUGAR) and another that said SODIUM CYANIDE – poison. After the subject affixed the labels to the bottles of his choosing, the experimenter spooned sugar from the two bottles into two plastic cups of water and asked the subject to rate how much he'd like to drink from each. Forty-one of fifty subjects opted for the sugar-labelled water – as if the word *poison* can actually poison water.

If we treat words and names as true extensions of what they represent, it may be in part because when we read a word, we automatically create a simulation of its meaning in our heads. For example, reading the words *lick*, *pick*, and *kick* activates areas of the motor cortex corresponding to the tongue, hand, and leg, respectively. So just to think about a word is to experience a bit of what it represents. This fact would explain the finding by Rozin that people will even feel uncomfortable drinking a beverage labelled NOT POISON. The word *poison* is enough to conjure dire images, and those images feel a little bit real.

Nowadays the name *Britney* is just as toxic as the word *cyanide*. Its popularity as a baby name peaked soon after Britney Spears released her debut single and has been on a downward trend ever since. The Harvard sociologist Stanley Lieberson has documented other instances of 'the symbolic contamination of names'. As Hitler rose to power, the name Adolph became taboo, and the name Donald went on a fifty-year decline beginning in 1934, the year of Donald Duck's debut toon. (If it sounds like a duck . . .) Maybe we just like or dislike certain names because they have positive or negative associations. Or maybe we fear that a child named Adolph contains some of the essence of the infamous Adolf. The hypothesis has not been tested. What *is* well established is that we gravitate towards people and things that share our own names or even just our first initials. We assume they are like us.

|||||||||||||||

In voodoo, placing the name of a victim on a doll or an effigy supposedly calls forth that victim's essence so you can then act on it. Explicit belief in the summoning power of names exists not only within sorcery circles but also within the walls of psychiatric hospitals. No surprise, right? If you're in a psychiatric ward, there's a good chance you'll have a number of irrational beliefs. But in a 1983 article in *Psychiatric Quarterly*, a clinical director of Worcester State Hospital in

Massachusetts reported the popularity of the name belief not among inpatients but among employees.

'I just remember sitting around with staff, chatting, and they had a number of magical ideas that they developed over time,' Wilfrid Pilette, the article's author, told me when I called him up. For example, don't mention the names of discharged patients, or they'll return. 'If it's a patient you like, if there was a strong investment and a strong connection', he said, 'there's the sense that you were protecting them by not saying their name.' (Staff tried not to jinx themselves in other ways too: 'One of the biggest things was you would never say no one's ever suicided here.') 'Anybody who works in health care is prone to that kind of thinking', Pilette says. 'Put under enough stress, you're going to get real magical.' He notes that much of their job is unpredictable. 'It's just another way of feeling some control that's not really there, it's just another defence mechanism, another way of alleviating anxiety.'

The conjuring potential of names plays out in numerous ways, from religion (many Jews will not say the name of God, YHWH, aloud) to dating and Harry Potter books (people occasionally refer to both exes and Lord Voldemort as 'He-Who-Must-Not-Be-Named').

Recited curses often employ not only names but taboo words, and taboo words have strong powers over us – capacities reflected in many national obscenities laws. Their power may derive in part from a sense that they summon taboo subject matter. The linguists Keith Allan and Kate Burridge wrote that 'taboo terms have been contaminated by the taboo concepts they represent. . . . This is why taboo words are often described as unpleasant or ugly-sounding and why they are miscalled dirty words.' The word itself becomes filthy.

We may see numbers as lucky or unlucky for the same reason we like our own names and dislike dirty words: we associate them with what they represent and attribute to them summoning powers. In the US and the UK, as well as in several other countries, thirteen is the source of the most common number superstitions. Triskaidekaphobia (fear of thirteen) and triskaidekaphilia (fondness for thirteen) have sprung up for various reasons throughout history, but the current trend can be blamed on the symbolism of Jesus and his twelve apostles dining together before his Crucifixion. Even if thirteen doesn't hold that particular association for you, it likely brings to mind some nebulous spectre of dark possibilities. In any case, most property developers still don't label the thirteenth floor of office and residential buildings floor number 13. High-rises are literal monuments to magical thinking. Further, some airlines don't label the thirteenth row as such. And in 2007, Brussels Airlines had to repaint its planes to add a fourteenth dot to its logo after complaints from passengers – as if the bad stuff represented by thirteen dots could somehow inflict itself on the functioning of a jetliner.

The number seven also holds significance across many cultures, and despite there being seven deadly sins, seven is somehow lucky. The attentive reader will notice that I have defined seven laws of magic for this book. Conceivably, I could have divvied the research into six or eight laws. Did I pick seven for logical, aesthetic, or superstitious reasons? Or merely as a cheap marketing ploy? I suppose the world will never know.

Oaths are about as close to functional magical spells as you can get. Say the right words in the right order and you change reality, though only social reality. While most language communicates information about the world, oaths *act* on the world.

Word pseudomagic is not merely a feature of the legal system; it is a basic function of natural language. The British philosopher of language J. L. Austin described certain acts of speech as *performatives*. In his 1962 volume *How to Do Things with Words* he gives these examples: 'I do' (at a wedding); 'I name this ship the *Queen Elizabeth*' (while smashing a bottle against the ship); 'I give and bequeath my watch to my brother' (in a will); and 'I bet you sixpence it will rain tomorrow.' As would be predicted by the law of similarity, cause resembles effect: the situation immediately conforms to the description of it. The hand is taken in marriage, the ship named, the watch bequeathed, the bargain entered. Even children on the playground use performative utterances quite adeptly. Witness: 'Dibs!' ('I declare this entity my property'); 'Fives!' ('I hereby render the seat I currently depart mine for taking upon my return, should that return come within five minutes'); and 'Uncle!' ('You shall cease committing acts of physical violence upon my person, for I surrender to your corporal dominance.')

The philosopher and literary critic Kenneth Burke has argued that belief in the magical power of symbols is based in everyday methods of communication. In 1950 he wrote that 'the realistic use of addressed language *to induce action in people* became the magical use of addressed language *to induce motion in things*'. So credence in the ability of words or symbolic actions to induce distant physical effects might have roots in the social use of performative utterances. We're expecting imitations or descriptions of a desired reality to work on the natural world just as they do on other people. We're anthropomorphizing the universe.

||||||||||||||

When Castignoli buried the Ortiz jersey, he was clearly hoping the Red Sox would haunt the Yankees' new haunt. But some fans suspected the curse may have backfired and the magic may have run in the opposite direction. Instead of an Ortiz essence flowing into the stadium, the significance of sepulture in the stadium may have flowed back towards Boston. In an online forum, a fan noted that Castignoli 'may have just buried his own team'. Or, specifically, buried Ortiz. A Red Sox fan on one site wrote: 'No wonder Ortiz is having the worst start of his career. 3/43 .070 average . . . It's because his jersey was in Yankee Stadium. I'm glad they found it cause now Ortiz will go back to his normal dominant self.' A Yankees fan pointed out ten days after the excavation that, over those ten days, Ortiz's average hits at bat was a somewhat dominant .306, to which another fan replied, 'There really was a curse!'

Voodoo Defeat

Putting faith in the law of similarity can be deadly. 'In records of anthropologists and others who have lived with primitive people in widely scattered parts of the world is the testimony that when subjected to spells or sorcery or the use of "black magic" men may be brought to death,' the anthropologist Walter Cannon wrote in 1942. 'Among the natives of South America and Africa, Australia, New Zealand, and the islands of the Pacific, as well as among the negroes of nearby Haiti, "voodoo" death has been reported by apparently competent observers.'

Cannon explains that whether or not one has actually suffered a supernatural slight, simple fear of death arouses a fight-or-flight response. When preparing to meet danger, adrenaline floods the body, constricting blood vessels and forcing blood into tissues and out of circulation; eventually, blood pressure can drop enough to lead to cardiac arrest. A 2002 retrospective in the *American Journal of Public Health* largely concurred with his hypothetical account and fortified it with more recent knowledge about stress hormones that could hasten a fatality. Voodoo death, of a sort, is real.

And voodoo may lead not only to death but to defeat on the baseball field.

On Friday, 5 October 1945, a businessman named William Sianis showed up at Chicago's Wrigley Field for game four of the World Series, between the Chicago Cubs and the Detroit Tigers, with two tickets in hand – one for him and one for his pet goat. Sianis, a Greek immigrant nicknamed 'Billy Goat' for his goatee, presumably hoped to promote his Billy Goat Tavern. The guests were booted before the end of the game, however, because of odour complaints. Here accounts differ, but Sianis is said to have cursed the team to never again win a League championship, or perhaps a World Series. He may also have telegrammed the losing Cubs owner, P. K. Wrigley, 'Who smells now?' From 1876 to 1945, the Cubs won the National League championship sixteen times. But they lost that game with the goat in partial attendance, along with the overall title, and have not since returned to the World Series.

Playing under a curse is similar to a phenomenon called *stereotype threat*. In 1995 Claude Steele and Joshua Aronson reported that black Stanford undergraduates perform worse on tests when told the tests are diagnostic of intelligence. The authors concluded that black students are aware of the stereotype that blacks aren't as smart as whites, and subtle reminders of the stereotype (even just being told they're taking an intelligence test) increase anxiety and hinder performance. Stereotype threat has been demonstrated in other minorities, in women, and even in white men on a physical task framed as an assessment of 'natural athletic ability'. As the American social psychologist Nicholas Herrera has pointed out, the Chicago Cubs have been stereotyped as the 'lovable losers', destined to mess up any chance they get of making it to the World Series; constant reminders of the curse keep them on edge, making the curse a self-fulfilling prophecy.

Their most poignant piece of self-destruction in recent memory came in 2003. Playing the Florida Marlins, the team was just five plays away from winning the National League championship and defeating the curse. Then the Bartman incident occurred.

Luis Castillo of the Florida Marlins hit a foul ball just over the edge of left field. As the Cubs outfielder Moises Alou reached into the stands to grab it and make the out, several fans unaware of his approach also attempted the catch. The ball deflected off the hand of one Cubs fan in particular, Steve Bartman, and eluded the embrace of Alou, who threw down his catching glove in anger. Alou told the *New York Times*, 'It wasn't the whole ballgame, but the Marlins had been playing with a lot of good luck, and we had bad luck. And when I was prevented from catching the ball, I think a lot of our players thought, "Here we go again."' The Cubs, of their own devices, went on to give up the lead and the game, as well as game seven and thus the series. Their championship dreams suffered voodoo death by stereotype threat, and Steve Bartman became a reincarnation of the original scapegoat.

Fortunately, the Chicago masses realized they could not string up Bartman, so they did the next best thing. The man in the stands behind Bartman who recovered the ball sold it for about $100,000 to Grant DePorter, the president of Harry Caray's Restaurant Group. DePorter hired a Hollywood special-effects expert to orchestrate a dramatic end for the artefact and, hopefully – symbolically – the curse. They placed their sacrificial-lamb-by-proxy in a clear box in front of an audience and blew it up.

The ritual didn't work, of course, but the story of the Bartman Ball gets more interesting after that. In homeopathy, you dilute a poison to nothing and use the remaining essence as medicine. DePorter's chefs took the remains of the poisonous ball, boiled it in Budweiser beer and vodka, and placed the condensed steam (which contained no residue of the actual ball) in a marinara sauce and served 'Foul Ball Spaghetti' to thousands of fans for $11.95 to cure them of their malaise. DePorter hoped to counter Sianis's sorcery with some warlock's brew of his own.

If Gino Castignoli's David Ortiz jersey hadn't been removed, it's possible such a well-publicized gimmick would have got into the heads of some of the players, creating stereotype threat of the voodoo variety.

But Castignoli says that even though the Yankees didn't know about the shirt in 2007, and even though they were still playing in their old stadium, the shirt did its job. 'I tried my hardest to curse them. It worked for a year. They got knocked out of the playoffs that year, first round. I really gave it to the Yankee fans.'

||||||||||||

The voodoo death of the Cubs' playoff dreams demonstrates one of the dangers of belief in the law of similarity, and something to guard against. A symbol, whether it's a curse or a burning flag or rain on an important occasion, is often treated as inherently efficacious and potentially damning. But these effects are mediated by our own expectations – expectations we have some ability to reconfigure.

The law of similarity also has positive uses. Holding a photo or a memento close can provide comfort or confidence. Symbols can be manipulated in the service of metaphor therapy. And the use of rituals and mantras and lucky numbers can make us feel we have some control over our woolly environs.

More broadly, understanding the essential role of similarity in how we think allows us to find and create physical metaphors that set the tone for the lives we would like to lead, from wearing a red shirt on a day you want to feel energized to rehearsing a scenario you want to become a reality. The true potency of symbols lies in their ability to alter our expectations and interpretations of the world. And our interpretations of the world are all we have – your consciousness is your reality – so, in a way, symbols do have magical power.

3

Actions Have Distant Consequences

Using Superstition to Make Luck Work for You

They say there are no atheists in foxholes. It has also been said that there are no atheists past Cape Omni – or whatever local cape separates a fisherman from the open sea. According to the US Bureau of Labor Statistics, commercial fishing is by far the deadliest job in America. It may also have the richest culture of rituals and taboos.

Sea lore contains a litany of rules to follow. Never bring a black suitcase or a banana onto a fishing vessel. Never whistle in the wheelhouse lest you whistle up a storm. Turn a hatch cover upside down and your boat might follow suit. Don't mention horses or hang mugs with the opening facing out. And good luck finding anyone in the Alaskan crab fleet – those grizzled seamen made famous by Discovery Channel's *Deadliest Catch* – who'll leave port on a Friday.

'We left town to do sea trials one time on the *Wizard* on a Friday', Keith Colburn, the captain and owner of the *Wizard* and a star of the programme, told me. He'd just wanted to test the boat out. No harm, as long as they weren't actually fishing. But another boat lost steering in the lock and rammed them, barely missing the fuel tanks. 'Just an absolute fluke', he said. 'At this point I do my best not to leave town on a Friday, I'll tell you that right now.' As for the rest of the fleet, 'Guys are leaving at 12:01 *all* the time.'

Mike Day, a fisherman who helps film *Deadliest Catch*, has left on Friday twice in his career. One time their boat was immediately caught in a surprise storm and returned to port encased in four feet of ice; the second time a rogue wave injured two men. 'That's one superstition I don't want to fuck with anymore.'

Colburn has a number of more individual superstitions. 'There's a multitude of things that I will and won't do when I'm fishing, just because.' He won't leave town without his lucky hat, lucky coffee mug, or the clipboard he's used for the last eighteen-plus years. For the last twelve-plus years he hasn't drawn the columns in his logbook. 'I've shut the boat down before to have either my brother or my mate Gary come up, draw the columns, and then off I can go.' And he surrounds himself with trinkets – bobblehead dolls, a hula girl, a lucky rock, a lucky wizard from his daughter, a rubber shark from his son. 'I've got a little shrine, actually, in the wheelhouse.'

Colburn isn't particularly mystic about his rituals. He knows the magic is all in his head. But he follows them anyway. 'If we leave any day of the week but Friday and we have a bad season or a breakdown or an emergency, we don't associate it with a particular day of the week. It's just something that happens', he says. 'But, in our brains, if we left on a Friday and something happened, we're gonna pin it on the day that we left.' I got the same story from other fishermen I talked to. 'I'm afraid to fly', one told me. 'I know deep down it's the safest way to travel. Same thing with the superstition. We know better, but we still go by it.'

Why would these salt-scabbed sea warriors, with their lives and livelihoods literally on the line, stray from pragmatism and invest in such whimsy, dancing with ghosts in gale winds? Who has time to tend to nagging compulsions about hanging a mug the right way or avoiding a taboo word when you're wrangling eight-hundred-pound steel cages amid heavy machinery on a rollicking deck in the frosty, frothy Bering Sea, where a six-minute dip in the drink means death? Is it possible that adherence to fanciful habits, even under duress, might provide some advantage? And is it possible that we all might have something to gain from superstitious behaviour?

Creatures of Habit

Superstitious rituals and taboos are attempts at control. Perform certain acts, avoid certain others, and the world is at your fingertips. You just need to find the right hidden strings to pull while sidestepping the trip wires.

There are three main routes to picking up a superstition, each of which may work in conjunction with the others.

The first is via the law of similarity: certain actions or events have an obvious symbolism to them. Whistling encourages storms because breath is like wind. Upside down hatch covers represent upside-down boats. Friday is the day Jesus was crucified.

The second is through social transmission. Either you observe someone else performing a ritual or someone instructs you to follow suit. If something appears to work for someone else, you don't want to mess with it. Why risk learning from your own mistakes when you can learn from someone else's?

Finally, we accumulate superstitions through simple conditioning. The modern study of superstitious rituals began in the 1940s, not with human subjects, but with birds. B. F. Skinner's famous 1948 paper '"Superstition" in the Pigeon' describes what happened when he placed hungry pigeons in cages with mechanical hoppers that automatically dished out treats every fifteen seconds. The birds would go about their business, doing normal pigeon stuff, and suddenly food would appear. Skinner observed that six of the eight birds developed repetitive behaviours. One spun in circles, another stuck its head in the corner, two bobbed their heads back and forth and sometimes took a few sidesteps as if performing their own dance routine. 'The experiment might be said to demonstrate a sort of superstition', Skinner wrote. 'The bird behaves as if there was a causal relation between its behavior and the presentation of food, although such a relation is lacking.' He added, 'There are many analogies in human behavior.' A great *New Yorker* cartoon by P. C. Vey has a businessman sitting at a computer saying, 'I don't get it – last time I jiggled the mouse this way we made a 16.45 percent profit.'

Follow-up experiments by other researchers called Skinner's interpretations into question. Skinner emphasized the differences between the pigeons' behaviours rather than the commonalties and failed to deduce that they were carrying out evolved foraging routines rather than independently learned 'superstitions'.

But Skinner's idea about accidental conditioning as a cause of superstition was essentially correct, as work with humans eventually revealed. Consider a 1987 Japanese study by Koichi Ono. Three levers sat on a table. On the wall behind them were a light and a number counter. Subjects were told, 'The experimenter does not require you to do anything specific. But if you do something, you may get points on the counter. Now try to get as many points as possible.' No matter what they did, once or twice per minute over the next half hour the light would flash red, green, or yellow; a buzzer would sound; and the counter would tick up.

Most subjects displayed 'transient superstitious behaviours'. That is, they went through periods where they adhered to an arbitrary pattern of movement such as pulling the levers in a particular order, thinking it was giving them points.

The description of one subject is worth quoting at length:

About 5 min into the session, a point delivery occurred after she had stopped pulling the lever temporarily and had put her right hand on the lever frame. This behavior was followed by a point delivery, after which she climbed on the table and put her right hand to the counter. Just as she did so, another point was delivered. Thereafter she began to touch many things in turn, such as the signal light, the screen, a nail on the screen, and the wall. About 10 min later, a point was delivered just as she jumped to the floor, and touching was replaced by jumping. After five jumps, a point was delivered when she jumped and touched the ceiling with her slipper in her hand. Jumping to touch the ceiling continued repeatedly and was followed by points until she stopped about 25 min into the session, perhaps because of fatigue.

Anyone who has ever dealt with a set-top TV antenna can certainly relate. You nudge the rabbit ears one way and you get fuzz. The other way and you get *The Price Is Right* – until you take a step back. You start dancing around and holding contorted positions trying to find the perfect reception, and pretty soon you're doing the macarena on the couch with your dog wrapped in tinfoil and the neighbourhood children holding hands in a circle around your house.

Sometimes we associate a success not with something we do but with an element of the environment, and we try to reproduce those conditions. 'If we happened to have a song on the tape recorder when we got into a big bite of salmon, we'd keep playing it, no matter how bad it was,' Bradford Matsen, the author of *Fishing Up North: Stories of Luck and Loss in Alaskan Waters*, told me. The Warren Zevon song 'Lawyers, Guns and Money' was on one morning when he and his partner were fishing Alaskan waters. 'We're forty miles off the coast with Mount Fairweather to the east, and it's sunrise and it's just absolutely stunning', he says. 'We get into a clatter of thirty-pound king salmon. They were almost cut from the same mould. We pulled until our arms almost fell out. We played that song for the next week, over and over and over.' Matsen says he's not alone with the habit of associating music with catch size: 'Guys would come in and say a particular band really fished well this trip.'

Clothing has the same effect. 'People have their lucky hat', another fisherman told me. 'If you got a crummy old hat and go out and plug the boat with halibut and make fifteen thousand bucks in a day, everything on you is your lucky thing. Your boots, socks, underwear.' Note, however, that if you keep your lucky underwear on for extended periods, you might hamper your chances of getting lucky upon returning home.

This kind of superstition is an example of what's called an *illusory correlation* – an overestimated relationship between two types of events, such as jumping up and down and receiving a point on a counter. We're not so great at gauging *actual* correlations without a pen and paper. Let's say some days you play a certain song and catch a lot of fish, and other days you don't play it and you come up empty-handed. Great. But on some days you play it and don't catch a thing, and on others you don't play it and you strike gold. The actual correlation is defined by the relative frequency of all four situations. It's not a calculation we perform intuitively. Instead, we tend to focus just on the hits – those days when the song and the full load overlap – and possibly overestimate the relationship.

Further, we often count near hits in the hits column. In one study by Thomas Gilovich, people were asked about the (relatively recent at the time) 1980 men's college basketball championship game. In the event, the University of Louisville edged out the University of California at Los Angeles thanks to a contested call made by a referee in the last two minutes. Over the phone, fans were asked to predict who would win in the case of a rematch. Louisville fans nearly all thought their team would win again, but the predictions of UCLA fans hinged on whether the interviewer reminded them of the fluke ending. Only 23 percent of those not reminded said UCLA would reverse course and win the second time around, but when the referee's contentious call was brought to mind, that number shot up to 70 percent. We take wins at face value but come up with excuses to think of almost wins as actual wins or should-have-been wins. Maybe you played 'Lawyers, Guns and Money', and you *would* have caught a boatload, but somehow your line broke, giving you a scapegoat and the chance to keep your theory about the song's effectiveness intact. Warren Zevon would have worked his magic if it weren't for that damn line.

Once we have a hypothesis about a correlation, we're not likely to look for disconfirming evidence. In one study, subjects had to judge whether an astrological personality profile matched a target individual. They asked the target more than twice as many questions soliciting information that would confirm the profile, versus questions that would evoke conflicting responses. We'll even selectively weed out evidence that directly counters our hypotheses. In a study where subjects read about scientific findings that either supported or dismissed the existence of ESP, most people correctly remembered the content of what they'd read, except for believers who read that ESP wasn't real, most of whom failed the recall test. People have a *confirmation bias:* once our minds are made up, it's hard to change them.

Illusory correlations between cause and effect can also take place at the perceptual level. 'Take, for example, the not-especially-uncommon event wherein you honk your car horn, and at just that moment a streetlight goes out', Brian Scholl, the director of Yale University's Perception and Cognition Laboratory, says. 'You may never for a moment *believe* that your honk caused the light to go out, but you will irresistibly *perceive* that causal relation. The fact remains that our visual systems refuse to "believe" in coincidences.'

The Belgian psychologist Albert Michotte wrote in 1945 that three factors contribute to a theory of causation: priority, exclusivity, and consistency. We will perceive that A causes B if it happens before B, if there are no other likely causes of B, and if A and B are conceptually related. Scholl's horn example meets the requirement of priority, sort of meets the requirement of exclusivity (a short circuit could also cause a light to go out), and might meet consistency depending on how you look at it (horns and streetlights are both traffic related). Research shows that consistency helps but is not strictly required for the judgement of causality. Apparently, we are just fine with the idea that actions can have surprising and counterintuitive effects.

Illusory correlations can leave you all wet whether you're judging which lures best hook fish or noting which pickup lines best snag dates or diagnosing which health habits best prevent catching colds. Thinking like a scientist – taking anecdotal evidence lightly, considering multiple explanations, looking for disconfirming evidence – doesn't always come naturally, but it's a good set of habits for navigating the world's uncertain waters, at sea or on land.

To Believe or Not to Believe

We often have difficulty judging when we have control and when we don't. The Harvard psychologist Ellen Langer (to be played by Jennifer Aniston in a biopic) has found that people most often feel an illusion of control in games of pure chance when the games involve features reminiscent of skill-based tasks. In one study, office workers purchased raffle tickets for $1 each. Half the workers selected their own ticket, and the other half were assigned tickets. When offered the chance to sell their tickets before the drawing, those with assigned tickets asked for an average of $1.96, but those who'd selected their own demanded $8.67 – more than four times as much as their coworkers. Merely selecting which ticket to buy induced the feeling that they had some control over the lottery and that they had exercised that control by choosing a ticket with a high chance of winning. These results explain why the popularity of lotteries took off in the US once states allowed players to select their own numbers.

Feelings of control frequently overstep their bounds in everyday life. 'The difference between skill and chance is often hard to delineate', Langer told me. 'Let's say you're playing poker and you won. Did you win because you're skilled or because you got good cards? If you're playing tennis, which seems like a totally skill activity, you have factors like the surface, the weather, whether your opponent slept well, and so on.' When there's any leeway in assigning causality, we tend to take credit for our successes.

Wagering that you have control when it's not clear that you do often makes practical sense. Consider this well-known wager: in *Pensées*, the seventeenth-century French mathematician Blaise Pascal wrote, ' "Either God is, or he is not." But to which view shall we be inclined?' Should you believe or not believe? Let's say you believe in God. If you're right, you go to heaven (big win); if you're wrong, you wasted some time in a pew (small loss). Let's say you reject God. If you're wrong, you miss out on an eternity of bliss (big loss), but hey, if you're right, at least you had your Sundays free for a while (small win). Pascal reasoned that unless the probability of God's existence is zero, you should put all bets on belief.

Humans make such wagers all the time without realizing it; they're called *cognitive biases*. I mentioned in this book's introduction that men overestimate the romantic interest of women so as not to miss a scoring opportunity (eternal life for his genes). Meanwhile, women overestimate men's caddishness to avoid getting played. According to the evolutionary psychologists Martie Haselton and Daniel Nettle, we're 'paranoid optimists'. We adhere to both 'nothing ventured, nothing gained' and 'better safe than sorry' in various realms of life. Haselton and David Buss explain such a mixed strategy with *error management theory:* if, over the course of human evolution, making a mistake in one direction hurt more than making a mistake in the other direction, we'll consistently veer towards making the less costly mistake, even if we end up making more mistakes overall than if we attempted a true course. Picture walking along the top of a low wall along a ledge. If you fall to the right, you plummet ten metres. If you fall to the left, you drop only one. I suspect you'll lean to the left a little.

That's why we startle at the sight of a stick on the ground that vaguely resembles a snake. The embarrassment of flinching at a branch (dropping a metre) pales in comparison to the danger of missing a real snake (splat). Our perception leans to the left.

The ever-handy error management theory may even help explain the so-called *Lake Wobegon effect*, where most people believe they're above average: inflated self-esteem gets you places; self-denigration does not. Better to try out for the team and fail than not to try out for a team you would have made.

And most important for our purposes, error management theory predicts the illusion of control. For years, whenever I stepped onto an aeroplane, I placed my hands on the outside of the fuselage. The chances were slim that my little ritual had any effect on keeping the plane's trajectory horizontal rather than vertical, but the cost was negligent and the potential benefit was great. It couldn't hurt, so why not? Better to overestimate the amount of influence you have on your surroundings for those cases when your actions turn out to have real effects.

So while any one instance of superstition may seem silly, the overall error management strategy that leads to superstition is supremely rational.

Stress Test

In times of stress, when there's little room for error, we tend to buckle down. One might expect that on a bobbing ship hundreds of kilometres from land a sailor's grip on reality would tighten. The Polish anthropologist Bronisław Malinowski found that even the tribal Trobriander Islanders in the South Pacific rigorously adhered to practical considerations on the water. 'They have, in fact, a whole system of principles of sailing, embodied in a complex and rich terminology, traditionally handed on and obeyed as rationally and consistently as is modern science by modern sailors', he wrote. 'How could they sail otherwise under eminently dangerous conditions in their frail primitive craft?'

But he found that these dangerous conditions, which forced the fishermen to control every element of the mission they could, also drove them to seek control over things they couldn't. 'Even with all their systematic knowledge, methodically applied, they are still at the mercy of powerful and incalculable tides, sudden gales during the monsoon season and unknown reefs', he wrote. 'And here comes in their magic, performed over the canoe during its construction, carried out at the beginning and in the course of expeditions and resorted to in moments of real danger.'

In contrast, Malinowski discovered that the safe and reliable practice of inner lagoon fishing featured very little magic. Only the dangerous and uncertain nature of open-sea fishing brought out the spells.

A study of New England fishermen in the 1980s echoed Malinowski's observations. John Poggie and Richard Pollnac of the University of Rhode Island interviewed 108 fishermen and asked them to list all the superstitions they were familiar with (a rough measure of how superstitious they were). The men also identified the average length of their trips. Poggie and Pollnac found that those who stayed out longer named more superstitions. Each day you're out there puts you farther from shore and deeper in danger. You're isolated from help in the case of injury or equipment failure and removed from safe waters in the case of a storm. (Fatigue and sensory deprivation from long spans on open water also break down rational inhibitions and lead to magical thinking.)

Superstition increases in the face of many types of stressors as a way to regain control. During the Gulf War (1990–91), Iraq fired forty-three Scud missiles at Israel, rendering many civilians scared for their lives on a daily basis. Giora Keinan of Tel Aviv University suspected the uncertainty might increase magical thinking, and so in February 1991 he gave a questionnaire to adults in four Israeli cities. Respondents rated their agreement with items grouped into three categories of magical thinking. The sentences represented the law of similarity (for example, 'If during a missile attack I had a photograph of Saddam Hussein with me, I would rip it to pieces'), the law of contagion ('I have the feeling that the chances of being hit during a missile attack are greater if a person whose house was attacked is present in the sealed room' – the room families gather in during attacks), and superstitious rituals ('To be on the safe side, it is best to step into the sealed room right foot first'). People also reported their tolerance of ambiguity by responding to items such as 'Before an examination, I feel much less anxious if I know how many questions there will be.' Keinan found that those who lived in cities at higher risk of attack (Tel Aviv and Ramat Gan) relied on all three forms of magical thinking more heavily than did those in low-risk cities (Jerusalem and Tiberias). People with a lower tolerance of ambiguity also thought more magically. Education level had no influence.

For some, maths exams can seem as stressful as missile attacks. In a paper titled 'Malinowski Goes to College', Jeffrey Rudski and Ashleigh Edwards documented the diversity and purported effectiveness of superstitions among a student population. (An earlier study found 70 percent of students reporting test-related superstitious practices.) The most common rituals were the wearing of particular clothing, knocking on wood, and avoiding saying things that might create jinxes. Students also reported the likelihood of using their favourite rituals or charms in various scenarios. For three types of events (a test, a dance performance, and an athletic competition), people were more likely to use their rituals and charms when they felt unprepared or when the outcome was important (a final exam or a championship game, say).

Daniel Albas and Cheryl Albas spent over a decade gathering data on hundreds of students' use of magic during exams. Some rituals obey the law of contagion: people wore jewellery owned by a bright parent or they used the pen they took lecture notes with because it 'knew' the answers. Some obeyed the law of similarity, such as a student's tradition of eating a sausage link and two eggs arranged on the plate to read '100' or stepping into the exam room with the right foot first to ensure right answers. Other rituals were arbitrary, such as the circling of the building before entering it or wearing a pink shirt. 'This is nonrational behaviour', they wrote, 'in a setting where one might expect maximum rationality.'

Along with fishermen, soldiers, students, and gamblers, professional athletes are one of the most superstitious groups on the planet, thanks to the level of competition and uncertainty in their work. In a famous essay titled 'Baseball Magic', the anthropologist and former first baseman for the Detroit Tigers George Gmelch wrote, 'As I listened to my professor describe the magical rituals of the Trobriand Islanders, it occurred to me that what these so-called primitive people did wasn't all that different from what my teammates and I did for luck and confidence at the ballpark.' He goes through a roster of rituals among players. Jim Ohms placed pennies in his jock strap after victories, which would clang against his cup as he ran the bases by the end of the season. Wade Boggs ate chicken before every game for seventeen years. Mike Hargrove performed so many little rituals while batting he became known as 'the human rain delay', Players also relied on the laws of similarity – they requested the jersey numbers of former stars – and contagion – they avoided using bats after poor hitters had used them.

Gmelch notes that while pitchers and batters are intimately familiar with chance – a millimetre or a millisecond can separate an out from a home run – fielders catch 98 percent of balls that come to them. 'In professional baseball, fielding is the equivalent of the inner lagoon while hitting and pitching are like the open sea', he wrote. As a result, 'I met only one player who had any ritual in connection with fielding, and he was an error-prone shortstop.'

Under Control

So anxiety increases the desire for control. Does the feeling of control reduce anxiety?

Ellen Langer and a colleague, Judith Rodin, tested the effects of a nominal level of control among a group of nursing home residents. People on one floor received a pep talk – not so much, 'Put your mind to it and you can do anything!' More along the lines of, 'This is your home to do with what you want. Rearrange the furniture. Walk around.

Watch TV if you want. Oh, and here's a plant.' They all received plants and were expected to take care of them. Residents on another floor heard a slightly different talk. Basically, 'This is your home and we want you to be happy so we're doing our best to take care of you. Oh, and here's a plant. We'll take care of that for you, too.'

Three weeks later, 71 percent of the second group had become more debilitated, but 93 percent of the empowered group had actually *improved*. They were happier, more active, and more mentally alert. They talked to staff, visited other residents, and attended films on their floor. Langer and Rodin followed up eighteen months later. The group encouraged to take responsibility for their lives were still happier and healthier. And here's the kicker: Only half as many had died as in the other group – 15 percent versus 30 percent.

After a year of imprisonment in two German concentration camps, the Austrian psychologist Bruno Bettelheim wrote that survival 'depended on one's ability to arrange to preserve some areas of independent action, to keep control of some important aspects of one's life despite an environment that seemed overwhelmed and total'. Without empowering pep talks and personal responsibilities of the kind Langer's subjects received, a nursing home can seem very much like a death camp.

Some engineers have pointedly built the illusion of control into our modern technological environment. A representative for Singapore Airlines told the magazine *IEEE Spectrum* that in-flight entertainment systems (IFEs) are 'as important as food' for their passengers, and not just to prevent boredom. You're strapped to a seat in a tube hurtling through the clouds with your life in the hands of a man you can't see, for stretches of up to eighteen hours. 'From a psychological perspective, we use the IFE system as a means of providing almost the illusion of control for the passenger', he said. 'If you can start, stop, pause, and rewind from a broad slate of options, it gives you something very specific to do.' Closer to home, try pressing the DOOR CLOSE button the next time you're in a lift. If the lift was built since the early

1990s, that button doesn't do anything. They put it there to give you a sense of control. As the journalist Nick Paumgarten wrote in a surprisingly compelling article about elevators in the *New Yorker*, 'That the door eventually closes reinforces [passengers'] belief in the button's power. It's a little like prayer. Elevator design is rooted in deception – to disguise not only the bare fact of the box hanging by ropes but also the tethering of tenants to a system over which they have no command.'

Several laboratory studies have also shown that the only thing worse than having your ears cut off is having no say over which ear goes first. Okay, that specific conclusion is yet to be confirmed, but it can be extrapolated from real findings. In one experiment, subjects were told they'd have to take a series of tests, either in a prescribed order or in an order of their choosing. They all escaped the experiment without lifting a pencil, but those who'd expected to have no say in the order of the tests had sweatier palms. In other research, subjects felt self-administered electric shocks to be less painful than ones they had no control over. (Yes, you read me correctly: self-administered electric shocks. For science!)

Out of the lab, Richard Sosis, an anthropologist at the University of Connecticut, interviewed women in Tzfat, Israel, in 2002–3. This was during the Second Intifada, a time of intense violence between Palestine and Israel. Sosis inquired about the women's religiosity, levels of worry, and recitation of psalms. Thirty-six percent of the secular group said reciting psalms could improve the *matzav* (roughly, the 'situation'). 'The data indicate that even atheists may turn towards ritual practice under conditions of stress', he wrote. And they benefited: among the secular women, those who relied on rituals didn't live in fear of terrorist attacks. They gained the peace of mind to ride buses, visit restaurants, and walk through large crowds. Sosis argues that the peace of mind came from a sense of control.

While most people overestimate the amount of control they have over the world, some evidence suggests that a subset of the population has an accurate portrait of their influence: those with depression. In one set of studies, subjects pressed a button and sometimes a green light would go on. For some subjects, the green light was occasionally controlled by the button, and for others there was no connection at all. The researchers found that depressed subjects estimated their control over the light *better* than nondepressed subjects did. They called the phenomenon *depressive realism.*

Some other researchers have called this 'sadder but wiser' phenomenon into question by arguing that the experimental methodology was too artificial (it's a subtle argument I won't go into here), but it's well accepted that whether or not depressed people's estimations of control are actually more realistic than others', their estimations are definitely lower.

Accepting one's powerlessness – even if it's real – can prove dangerous. Numerous studies have shown that when animals suffer inescapable electric shocks – I know, I know, science can be so *mean* – the animals then make little effort to avoid easily escapable shocks. Similarly, humans have shown a reluctance to shut off loud tones blasted through headphones after experiencing a series of uncontrollable noises.

The psychologist Martin Seligman attributes this pattern of behaviour to *learned helplessness:* one becomes defeated and expects that future endeavours will likewise be futile. You believe there's no correlation between what you do and what happens to you. A sense of helplessness can become global, extending beyond the domain in which it formed to other areas of life: if your boss doesn't listen to your suggestions at work, you might call it quits on your diet. People in abusive relationships often learn not to stand up for themselves and become pushovers in general. It's easier than fighting back with futility. If one concludes that bad events are unavoidable, learned helplessness can contribute to or even cause depression. You look at existence as a series of hardships and you've just got to suck it up. You stop trying at life.

But why is it that some uncontrollable situations lead to learned helplessness while others lead to the illusion of control? Why do we sometimes give up hope and other times perform strange rituals? The psychologist Helena Matute identified a key factor: feedback. When subjects in her study tried to influence a series of outcomes and were continually rebuffed, they showed signs of learned helplessness: they performed worse on a follow-up cognitive task than subjects whose previous attempts at control had been successful. But withholding negative feedback from them allowed them to develop unchecked superstitious behaviours that they believed actually worked, and afterwards they performed just as well on the cognitive task as the group who'd had real control. Superstition protected them against learned helplessness. No reality check, no problem.

Matute's findings present a complicated picture of superstition's potential benefits. On the one hand, putting faith in your rituals and shutting out any disconfirming feedback can inspire a spreading self-confidence. On the other hand, no one would ever suggest ignoring feedback as a blanket policy; we learn from our mistakes. I hedge that a good middle road might be to use whatever rituals feel effective but always to remain on the lookout for something that might work even better.

Do You Feel Lucky?

Does reliance on superstitious ritual fit the mind-matter-meld definition of magical thinking? That description of magical thinking, you'll recall, involves treating the nonmental as if it had mental properties or vice versa – it's a confluence of the subjective and objective. (Certainly, symbolic rituals adhere to that definition, as I explained in the last chapter.)

A common definition of superstition is that it's a mistaken idea about causality. So if you cross your fingers and expect it to influence a raffle drawing, that's superstition, but is it magical thinking? You're linking one physical process – a gesture of the hand – with another physical process – a number being drawn from a container. In a way it's no different from pressing a button on a remote control to change the TV channel. Where's the involvement of the mental?

Perhaps the essential element behind a magical ritual is not the physical action but the intent behind it. I often knock on wood, but with no wood around I'll find a plant-fibre-based substitute (paper, cotton). It's the thought that counts, right? Some people even just say, 'Knock on wood', or think it to themselves. So perhaps it's mind-over-matter magic (explored in full in the next chapter). The ritual or charm just acts as an antenna for intention, focusing thoughts and sending them out.

Or maybe we believe rituals channel something else: luck.

Originally, psychologists saw belief in luck as a sign of giving up, of handing over the reins to external forces. But luck is also sometimes treated as either a personal attribute or a force one can wield skilfully, and research over the past fifteen years has tended to treat belief in luck – especially belief in good luck – as something to nurture. Maia Young of UCLA administered to people both a belief-in-good-luck scale – subjects rated their agreement with items such as 'I consider myself to be a lucky person' and 'Luck works in my favor' – and an achievement-motivation scale – 'Once I undertake a task, I persist', 'I prefer to work in situations that require a high level of skill'. She and her collaborators found that those who believed they were lucky had no intention of sitting back and letting good things come to them; they wanted to go out there and take on the world, with luck as their copilot. 'The more you think of luck as a stable, personal trait', she told me, 'the more you feel personal agency, and the more you have a preference for challenging tasks.' The idea is that if you think you're the type of person who goes into a difficult situation and comes out ahead, you're going to put yourself in more of them, and you're not going to accept defeat. People think of luck as a deployable skill.

In chance situations, belief in luck is actually what's leading to the illusion of control, according to research by Michael Wohl and Michael Enzle of the University of Alberta. And they demonstrated how the deployment of personal luck can work alongside the two laws of sympathetic magic. People selected a Ping-Pong ball with a number on it and then watched an experimenter spin a wheel of fortune. If the wheel landed on their number, they won money. Some of the subjects held on to their ball, and others were asked to set it aside. During each spin, they reported their chance of winning and their level of personal luck. Wohl and Enzle found that those who physically held their ball were more confident in winning, and that this was due to feeling luckier. So touching the ball conducts one's luck into it via the law of contagion, and the ball is connected to its corresponding spot on the wheel via the law of similarity. This explains why gamblers at the racetrack some-times kiss their tickets.

The feeling of luck you get by performing little rituals increases confidence, but can that feeling boost actual performance in mental or physical tasks? To find out, the psychologist Lysann Damisch per-formed a series of experiments at the University of Cologne. In the first experiment, subjects performed ten golf putts. When handing the golfer the ball, she said either, 'Here is your ball. So far it has turned out to be a lucky ball', or simply, 'This is the ball everyone has used so far.' Those with the 'lucky' ball made 6.42 putts on average, versus 4.75. That's a 35 percent improvement.

In the second experiment, subjects were tested on motor dexterity by manipulating a clear plastic cube to roll thirty-six little balls through a set of holes inside it. When Damisch prepped them by saying, 'Ich drück dir die Daumen' ('I press the thumbs for you', the German equiva-lent of 'I'm crossing my fingers'), subjects took a little over three minutes on average, versus about five and a half minutes for the other subjects.

For the third experiment, people were invited to the lab for a mem-ory test and asked to bring a personal lucky charm. Half of them kept the charm with them, and for the other half the charm was taken to another room. The subjects reported their levels of anxiety and of self-efficacy (their estimated ability to master the test) before trying to match eighteen pairs of overturned cards by looking at two at a time. The 'charmed' group performed better, a difference accounted for by their greater confidence.

Finally, experiment three was repeated with another group, with slight differences. For their task, they had to create as many words as possible using a string of eight letters in front of them. The subjects holding their rabbits' feet and whatnot persisted longer than the other group (about twelve minutes versus seven) and also found about 50 percent more words. Statistical analysis showed that their better performance was due to greater self-efficacy.

Self-confidence improves performance through several routes. People take on larger challenges and thus achieve greater accomplishments (as long as they don't set their goals too high). They're more persistent in chasing those goals. And they believe more factors in their success are under their control. So, for example, if they missed some early putts in experiment one, they may have taken that as constructive feedback about how to modify their swing and improve.

Previous research has shown that athletic routines (such as bouncing a tennis ball three times before a serve) can boost performance, but Damisch and her collaborators note differences between those routines and superstitions. First, routine movements work by focusing attention and by preparing particular motor sequences, not by increasing self-efficacy. And second, magical superstitions have an additional layer of meaning – they're associated with luck, and they're often symbolic.

How do people judge when rituals are effective? The research on illusory correlations suggests that we notice instances of using them and witnessing an intended effect. You crush your tennis opponent whenever you wear a particular shirt, so you draw a connection between wins and wardrobe. But Damisch's experiments point to another mechanism. After several of the tasks, people rated their performance on a 1–9 scale. The 'lucky' participants didn't think they did any better than the unlucky ones. The fact that they're not noticing differences in outcomes suggests that to them the more apparent effect of rituals and charms is the boost in confidence. This feel-good state acts as the prime reinforcement of carrying a lucky clover.

Laugh all you want at Wade Boggs for eating chicken before every baseball game – he even told the *New York Times*, 'Sometimes it's better to be lucky than good' – but he got the job done. Damisch wrote in her PhD thesis, 'in addition to simply increasing the acceptance of superstitions, we might also wish to systematically embed superstitions in everyone's life.' Typically, I support separation of superstition and state, but I see potential in her suggestion that 'it might be plausible to literally teach children the use of good-luck-related thoughts or behaviors.' No charm left behind. A horseshoe in every classroom. Counter to the saying about horseshoes working even if you don't believe in them, it turns out they work *because* you believe in them.

Yet it's possible to have too much belief in good luck. In an episode of the TV sitcom *30 Rock*, the ever-chipper character Kenneth tosses his wallet out the window to make a point. 'I'm not worried', he says. 'Why? Because everything always works out for the best.' The wallet actually does come back to him (maybe because of the 'lucky rabbit spine' tucked inside?), but that's hardly model behaviour. A lucky charm or ritual is no substitute for proper caution or preparation. (Barack Obama played basketball on the day of every primary in the 2008 presidential election once he linked the habit with success, but he didn't let up on the handshaking.) One study showed superstition's effect on risk taking in the context of a multiple-choice test. Among participants with high belief in their own good luck but low ability in the subject area, a fortunate event right before the test increased their confidence and their rate of guessing on answers, even though they got most of them wrong and there was a penalty for guessing incorrectly.

Overconfidence – produced by feelings of luck or otherwise – can have expensive consequences in the real world. There are obvious examples – a novice snowboarder shouldn't attempt a Double McTwist 1260 – and more subtle and systemic ones, such as audacity in investing. When researchers looked at 107 traders in four London banks, those traders with the highest illusion of control in a computer task were the least successful at their jobs. (That doesn't mean one can't profit from *other* people's superstition. Hot tip: stocks reliably dip in the three days surrounding solar and lunar eclipses and then bounce back.) While writing this chapter and thinking about luck, I started to feel lucky and bought a lottery ticket for the first time ever – another predictable way to lose money. Of course I knew the odds, but I persisted with the irrational belief that it was a wise investment. (Does that qualify this as a stunt book? *The Day of Thinking Magically*.)

A touch of luck might help you fish or putt or campaign, but it can't make you (or your wallet) fly. Even true-blue sorcerers understand the limits of their craft. The anthropologist Meyer Fortes once offered an African rainmaker a large fee for his services. In reply, Fortes was told, 'Don't be a fool. Whoever makes a rainmaking ceremony in the dry season?'

Obsessed

Even if lucky rituals don't encourage dangerous levels of risk taking, the mere necessity of their execution can hamper one's style. Keith Colburn, of the Wizard, always uses a Cup Noodles cup as a spittoon. 'Years ago I had a phenomenal season with this cup', he says. 'The next season I didn't have it, and I was doing horribly. I managed to get one from my brother on a different boat, and immediately my numbers improved dramatically.' He says there have been about three times he's had to retrieve a cup from another boat in the neighbourhood. 'As soon as I get it, boom, I'm off and running again.'

Gladden Schrock, a retired Maine fisherman and theatre professor, e-mailed me several superstitions, including this one (and yes, all his e-mails read like this):

Absolutely never say the word p-i-g on board a fishing boat. This would/will have immediate personal consequence, as you'd palpably feel the chill from the crew, or (extreme cases) the cap'n turns around whole-hog and heads home & ties up. About fifteen years ago there was a turf-tiff between lobster-men and purse seiners. One guy in particular was a pain in the ass; and someone took to breaking in on his CB, chanting 'Blue Pig, Blue Pig' . . . and drove the guy dead-heat home, all balmy-nutted & eye-popped, his thirty-six-footer showing a bone in its teeth the whole way. I was at the dock when he landed, the Hounds of Hell on his neck, looked like.

Children are especially confined by their own magical rituals. 'When I was a kid the bows on my shoelaces had to be the same length or nothing would go right', a friend told me. 'The battle for Middle Earth would be lost if you didn't put your clothes on in the right order. That's how intense it felt.'

Most children know the phrase *Step on a crack, break your mother's back*. A case study reported in the journal *Neurocase* describes a British boy with his own version: *Fail to step on a line, cause 9/11*. He suffered from Tourette's syndrome and obsessive-compulsive disorder. Among his compulsions was the need to step correctly on a particular white mark on the road every day. On 9/11, he forgot his duty. The boy's parents brought him to the clinic two weeks later. He was racked with guilt. Fortunately, he found relief in medication, as well as a reminder from his mother of the time difference – he'd missed his mark after the morning attacks.

I had my own quirky rules growing up. If I brushed against something on one side of my body, I had to reproduce the same sensation on the other side. I heard it was bad luck to sleep with your wardrobe door open, so I needed it closed every night. And the TV volume always had to be set at an even number. I've since found more important things to worry about. For example making eye contact while clinking glasses after a toast to avoid seven years of bad sex.

Reliance on superstitious ritual is a prominent part of obsessive-compulsive disorder. 'In OCD, the disorder has seemingly hijacked the human tendency to think superstitiously and run away with it', says

Fred Penzel, an American psychologist who has treated patients with OCD for thirty years and has written about OCD and magical thinking. Those with OCD suffer intrusive thoughts – fears that something bad will happen to them or a loved one, worries about physical contamination, concerns about general disorder in the world – and they feel compelled to perform certain actions to alleviate their anxiety. 'One of the hallmarks of OCD is pathological doubt that can be so strong, only magic will suffice to deal with it', Penzel says. Sometimes the routines become debilitating. Penzel describes one patient whose rituals for entering his own home became so complicated that he gave up and lived in the back of his car for a year.

Most of us aren't in danger of setting up camp in our cars, but superstitions can still get in the way of daily life. In the study of university exam takers by Albas and Albas, one student insisted on wearing a three-piece suit that had served him well on a job interview, no matter how well heated the exam room was. Another student couldn't take an exam until he'd found a coin on the ground, even if it meant sacrificing last-minute cramming for bus-stop scrounging.

Another woman had to eat a carrot muffin on the way to tests. Sounds like a good short-term energy source, but to not find one constituted an ill omen. If feeling lucky can improve performance, it's possible the anxiety provoked by missing a ritual could hinder it. So beware that if you invest too much in a prescribed routine, it may come back to haunt you.

Jinxed

Jen Castle of Ketchikan, Alaska, had fished on a few boats before she met her husband, but she had to prove herself before he would let her join his team. She'd go out with him for a week at a time, but she really wanted to be on his summer salmon crew. After three years she earned her spot.

That first summer, they were about a day out from the shore when Castle exclaimed to her brother-in-law how excited she was. 'I told him, "Take a picture of my teeth in case it's really rough and I get my teeth knocked out, ha ha ha."' As Anton Chekhov would say, the gun was on the table. 'Then midway through the trip I had a really bizarre accident.' Of course.

The wind was dead, waters were calm, the conditions were perfect. They weren't even fishing, just hauling the net from one part of the boat to another. 'I was doing something I have done many many times', she says. Then a large metal ring attached to the net system swung over and hit Castle in the mouth, splitting her lip open. Blood gushed everywhere. A nearby charter plane had to pick her up and take her to shore to get sewn up.

'Fisherman don't ever talk about, 'God, I hope that storm that's coming doesn't sweep us off the deck'", Castle says. 'I came from the corporate world, and people are always talking about scenarios. It helps prepare you for anything that could happen. But in fishing you never do it because people think you could bring a bad scenario onto yourself. After that experience I definitely felt that way because that was just too creepy.'

‖‖‖‖‖‖‖‖‖‖‖

There are certain laws of nature everyone accepts. The surest way to bring about rain on an overcast day is to leave your house without an umbrella. Is your queue at the supermarket moving too slowly? Switch queues. That will definitely speed it up (minus you). And if you've hit a series of green traffic lights that just might get you to the post office before it closes, comment on your string of success. Ah, there's the red.

Do people really believe such actions can change their fortunes?

In recent years the American psychologists Jane Risen and Thomas Gilovich have shed more light than anyone on the phenomenon of *tempting fate*. When they asked people to answer rationally whether exchanging a lottery ticket for another ticket would increase the chances of their old ticket winning, 90 percent said no. But when asked to answer the same question using their gut, 46 percent said yes. And when people imagined either keeping a lottery ticket or selling it to a friend, a stranger, or an enemy, they felt selling the ticket to an enemy gave it the best chance of winning, followed by selling it to a stranger, selling it to a friend, and, finally, keeping it.

In another experiment, people pictured a student applying to Stanford University and then receiving a Stanford shirt from his mother. They said wearing the shirt would make rejection more likely than stuffing it in a drawer would. (Risen can relate. When she interviewed at Stanford for a job they gave her a Stanford notepad, folder, and pen. 'I had this moment of thinking about whether or not I should take them with me', she recalls. 'Would it be presumptuous to take them while I was waiting to hear or rude to leave them behind? Instead of making me more rational, the research has made the magical beliefs especially accessible.')

Risen and Gilovich argue that belief in tempting fate rests, in part, on a three-step mental process.

First, some behaviours make outcomes seem especially bad because they highlight the contrast between what happened and what almost happened. Being stuck in a slow queue feels worse if you switched into that queue than if you were always in that queue, because if you switched, you'll be thinking about how you were *just in* a faster queue.

Second, negative scenarios engage our imagination more than positive ones; in general, bad things in the world can harm us more than good things can help us (a fish can feed a man for a day, but a blowfish can kill him for a lifetime), so we want to keep them on our sonar. Further, discomfort forms the basis of self-regulation; if something's wrong, we fix it. So we've evolved to attend more to negative than positive stimuli. (Our inflated focus on the bad plays out in the way we detect threatening faces faster than pleasant ones, in the way a single sickening bender can put you off tequila for years, and in the advice that a healthy relationship demands five compliments for every critique.) So if you're thinking about switching queues, the thought of switching to a queue that then slows down is worse than the thought of staying in a slow queue, and therefore it looms larger in your head.

Finally, the more you think about something, the more likely it seems, because of the availability heuristic.

To summarize the three-step process, negative outcomes feel worse after tempting fate, which makes them especially attention-grabbing and thus more likely seeming. Sounds like a rickety series of cognitive contraptions requiring a lot of effort to execute, but it's completely automatic. In fact, Risen and Gilovich found that asking subjects to count backwards by threes from 564 – a cognition-hogging task – made them *more* likely to believe showing up to class without doing the reading would get them called on.

What exactly do we mean when we talk about *tempting fate*? The phrase appeared around 1700, and *to tempt God* goes back to the 1300s. Risen and Gilovich had people sort fifty random newspaper articles that used the term *tempting fate* and found two dominant themes: unnecessary risk taking and hubris. Attempting to 'cheat death' or showing presumption about success will inevitably invite rebuke. As a proverb says, 'If you want to hear God laugh, tell him your plans.'

In the autumn of 2001, a man posted several letters containing anthrax spores to New York, Florida, and Washington, DC, creating a scare across America. A reporter for the *Washington Post* newspaper called up Scott Ian, a guitarist for the thrash-metal band Anthrax. 'People keep coming up to me and saying, "Hey, wouldn't it be funny if you got anthrax?"' he told the reporter. 'I'm like, "Oh, that'd be hilarious."' To be on the safe side, he filled a prescription for the antibiotic Cipro. 'I will not die an ironic death', he said.

'The universe seems interested not only in punishing certain behaviors but in punishing them a certain, ironic way', Risen and Gilovich wrote. We predict that negative outcomes will share some association with their antecedent – they'll fit the crime. Therefore, wearing a Stanford T-shirt will have no effect on the weather, and carrying an umbrella will have no effect on school admissions. And naming your act Anthrax offers no reason to stock up on, say, Minoxidil. (Naming your thrash-metal band Male Pattern Baldness, on the other hand . . .)

A predictable way to invite failure is to call attention to success. In one study, 48 percent of training doctors avoided the word *quiet* while on call for fear that all hell would break loose. You can even jinx other people this way. Some families in Azerbaijan seclude infants and their mothers for forty days after birth to make sure no one compliments the parents. In a study by Risen and Gilovich, an experimenter commenting on a subject's streak of success in a gambling game doubled the chance that he'd stop gambling at that point and take his winnings. He figured the jig was up.

Risen and Gilovich argue that thinking about the positive (a streak) automatically calls to mind its flip side (the streak's end), which then takes mental priority and thus seems more likely. They also suggest the superstition about calling attention to success results in part from a failure to appreciate *regression to the mean*. Any streak of success will eventually end, whether you note the streak or not. But remarking on it sticks out as an identifiable event, whereas keeping silent does not, and it's easier to draw a connection between the end of a streak and an identifiable event than a nonevent.

It's interesting that some people fear commenting not only on the expectation of positive outcomes but on the possibility of disaster, like those fishermen keeping mum about a storm. Paratroopers share a similar taboo, according to Col. Thomas Kolditz, the head of the Department of Behavioral Sciences and Leadership at the US Military Academy at West Point. The most common term for a skydiver hitting the ground is 'to go in', he says, a phrase that also comes up a lot in everyday speech. 'Among skydivers, anytime you inadvertently use the term "go in", someone will immediately stop you and say, "Don't say that."'

Presumably, jinxing yourself by discussing negative scenarios works in a similar fashion as jinxing yourself by discussing positive ones. Talking about getting washed overboard or having your teeth knocked out or hitting the ground makes those scenarios salient, and then the availability heuristic takes effect and they seem more likely. Unfortunately, this puts you in a predicament where talking about best-case scenarios and worst-case scenarios both screw you over, leaving discussion of any impending events taboo. 'When things are going good, try not to talk about it too much, and when things are going bad, certainly don't talk about it too much', Keith Colburn says. 'Either way there's going to be a storm.' So fear of jinxes can definitely get in the way of planning for the future.

But once you've tempted fate, there's at least one way to make amends, to tell fate, *Really, I take it all back*. Giora Keinan asked subjects a series of questions including 'Has anyone in your immediate family suffered from lung cancer?' 'Is your health alright on the whole?' and 'Have you ever been involved in a fatal road accident?' Half the participants spontaneously knocked on wood at least once after answering.

Purchasing insurance is another way people attempt to deter fate and reduce the likelihood of an accident. Orit Tykocinski of the Interdisciplinary Center in Israel asked people to imagine going on a trip to Bangkok. One group pictured their mother buying travel insurance for them, and another group pictured getting on the plane without enough time to purchase insurance. The insured group rated the probability of illness or lost luggage lower than the uninsured group did. Tykocinski argues that thinking about one's insurance provides an overall sense of safety that makes negative hypothetical events less threatening and thus seem less probable. (She notes that insurance agents 'upsell' by mentioning catastrophes not covered by basic plans. 'Refusing the added coverage at this point may be conceived of as an act that tempts fate', she wrote. 'All of a sudden, the possible risk of being hit by a tornado while vacationing in Rome seems plausible.')

There's a drawback to believing one can deter fate: carelessness.

It's one thing to believe that wearing a seatbelt will make a car accident less injurious or that purchasing a fire extinguisher will make a grease fire less damaging. But Tykocinski's data suggest we believe wearing seatbelts and owning fire extinguishers make disasters not just less harmful but also less likely. In reality, these protections tempt not fate but us – to behave recklessly.

Risen and Gilovich note that going against any superstition, no matter how arbitrary, from walking under a ladder to commenting on a winning streak, can tempt fate: 'To suffer a bad fate after flouting conventional wisdom and prevailing norms is especially painful because of the additional regret and embarrassment one feels for having unwisely gone out on a limb.'

Allowing anticipated regret to influence your judgements of objective probability is irrational, but the psychologists Dale Miller and Brian Taylor note that it's *not* irrational to let anticipated regret affect your decisions. Regret is real. So if you'd feel worse falling into a pothole after walking under a ladder than after walking around it, by all means, take the few extra steps to walk around. You won't regret it.

Overboard

'Everybody to some degree is superstitious', Colburn says. 'Some people say, "I am not superstitious, unequivocally, no way", yet they have their little regimented way they do things.' They tie their right shoe first or brush their hair a certain way or put their toothbrush back in the cup in a certain manner. 'Is that a routine or is that a superstition? They'll say it's just a habit, but if you told them to do it a different way, within two or three days if they have a bad day at work or they get into a fender bender, I'll be willing to bet that they go back into their "routine".'

And what if your bad day at work comes on the high seas? Hollis Jennings, who works the waters of California, Washington, and Alaska, shared a journal entry with me describing what happened in 2007 when she brought a suitcase aboard – a big no-no. She'd planned on making fun of her skipper for his apprehension about old stories, but, according to her journal, 'The voodoo soon caught up with us.' Over just a few days, the battery on the skiff dies, the net snags and tears (losing them thirty tonnes of squid), the purse line breaks, the skiff loses a cleat and starts drifting away, the boat's anchor sets itself, and finally a pump stops and the skiff fills with water. In order to make sense of their misfortune, and to regain control over their own fate, the crew heap blame upon, and take action against, the only reasonable culprit. 'Needless to say Will and I have just thrown the black suitcase overboard', Jennings writes. And just to clear up any bad juju, they plan on attending church the following Sunday.

'I feel obligated to mention that, for reasons not worth explaining, we have also been out of cigarettes since Tuesday', Jennings's entry concludes. 'Whoever said "a bad day of fishing is better than a good day at work" has never had a really bad day of fishing, much less a bad week.'

4

||||||||||||||||||||||||||

The Mind Knows No Bounds

Psychokinesis, ESP, and Transcendence

Football has 'the twelfth man' – the home side, whose spirited participation supposedly aids the efforts of the eleven players on the field. The three spaceborne astronauts of Apollo 13 had a 'fourth man' – one numbering over a billion. But does cheerleading do any good when the team is at an (extremely) away game and they can't hear you scream? Can you influence the outcome of a space mission simply by wishing for the best from afar? Presumably, millions believed they did.

On 13 April 13 1970, fifty-five hours and forty-six minutes after liftoff, Jim Lovell was wrapping up a TV broadcast: 'This is the crew of Apollo 13 wishing everybody there a nice evening, and we're just about ready to close out our inspection of [the lunar module] and get back for a pleasant evening in [the command module]. Good night.' Nine minutes later they heard a bang. One of the two oxygen tanks on the service module had exploded. The second was leaking. Two hundred thousand miles from home, the command module pilot, John Swigert, radios to Mission Control, 'Okay, Houston, we've had a problem here.'

Apollos 11 and 12 had been executed without major problems, and the series of moon shots had become stale in its success (a Milan headline described the Apollo program as 'Too Perfect: The Public Is Getting Bored'). But over the next four days, it once again induced nail-biting. Thousands of churches and synagogues held special services. The Chicago Board of Trade paused trading for a moment of tribute. On 14 April, the House and Senate passed resolutions asking businesses to pause at 9 p.m. to 'permit persons to join in prayer for the safety of the astronauts'. Millions of people kept abreast of the men's fates via TV and radio.

The concern was worldwide. People prayed at the Wailing Wall in Jerusalem. Pope Paul told a crowd of ten thousand at St Peter's Basilica, 'We cannot forget at this moment the lot of the astronauts of Apollo 13.' A commemorative Apollo 13 medallion shows two hands together and reads, 'And the whole world prayed.'

For their part, the people of NASA were doing what they could to ensure a safe splashdown. 'Physics really wasn't with us on this', Edgar Mitchell, who worked in the control room to help engineer the rescue, before walking on the moon himself on Apollo 14, told me. 'We had to use all the tools and all the skills engendered in that whole team.'

With the command module losing power, the crew piled into the lunar module, the craft meant to drop two moonwalkers down to the lunar surface and return them to lunar orbit (a goal they'd since abandoned). Instead of supporting two men for two days, the lunar module would need to keep three men alive for four days as they swung around the moon and returned to Earth. The team found ways to conserve power and water but risked dying from carbon dioxide poisoning until someone on the ground figured out how to modify the air filter using duct tape and cardboard. And without enough power for computer navigation, they had to find a way to align their trajectory with Earth using a handheld sextant.

'The chance of bringing back Apollo 13 was really a long shot for all of us, we knew it was a long shot, and many of us attributed the success to the whole world pulling for us', Mitchell says. Thoughts saving the day. He notes that 'of course there wasn't any strong evidence – there weren't any protocols by which you could say that – but it was just an intuitive feeling. It certainly didn't hurt to have all the good wishes of human civilization on our side.' Scientists and engineers feeling that well wishes might have saved three men doomed by the physics of their situation is a dramatic challenge to the idea that the logical mind is immune to magical thinking.

We all believe in mind over matter to some degree. Who hasn't made a wish while blowing out birthday candles? Resisted certain thoughts for fear of jinxing oneself? Tried to squeeze a penalty inside the posts using nothing but hope and concentration?

Believing you can control matter with your mind is but one example of the idea that the skull is a permeable membrane between our minds and the world. We also think we can sense distant events or the thoughts of others. And in special circumstances we can feel at one with the whole universe. Even though the brain is finite and completely contained inside the head, we can't help believing that the *mind* knows no bounds. It may be unknowable whether the mind has such powers – the question is, Why does the mind think it does?

It's the Thought That Counts

Prayer relies on one form of magical thinking – belief in a supernatural agent who answers appeals – but one can also believe in unmediated mental causation, the direct influence of mind over matter. Aside from all the prayerful entreaties launched to heaven in April 1970, no doubt plenty of atheists sent their hopes skyward, picturing a repaired spaceship bringing its crew home safely. And according to a set of clever studies by Emily Pronin and collaborators at Princeton and Harvard universities, just those thoughts about a rescue would have been enough to make the thinkers feel they'd participated.

The Princeton men's basketball team was attempting to achieve its sixteenth straight home-game win against Harvard, while also avenging a road loss against their rivals from earlier that season. It wasn't the finals, but it was high stakes for the two teams. Before the game, experimenters approached fans in the stands, handing them instructions and a two-page survey. The first page included photos of the top Princeton players along with facts about them. Half the participants were instructed to write down how each player could contribute to the game. Meanwhile, the control group just wrote down physical descriptions of the players. Later in the game, they all turned the page and rated how much they felt they'd affected the team's playing. Those who'd thought specifically about the players' performances beforehand felt more responsible for them than the control group did. The 'sixth man' was taking credit.

Princeton won that game 66–44, but would the fans have felt responsible for a loss? And would terrestrial spectators have felt any guilt about Apollo 13 had the recovery not gone so well? Pronin and her collaborators conducted another experiment during Super Bowl XXXIX, in 2005. Towards the end of the game, in which the New England Patriots beat the Philadelphia Eagles 24–21, experimenters handed surveys to people watching on the giant TV in the Princeton student centre. The survey asked how much they'd thought about the game (e.g., 'During this Super Bowl, how often did you think in advance about whether the upcoming play would be a run or a pass?'), how much influence they'd had on the game, and which team they were rooting for. As predicted, the more the students had thought about the game, the more responsible they felt for the outcome. And fans of the losing team felt just as accountable as winning fans, a fact indicating that taking credit for something you've thought about isn't just a self-serving bias. Surely, you've felt you jinxed yourself by thinking too much about an event before it turned sour.

It should also be noted that the game was played in Jacksonville, Florida, nearly a thousand miles from Princeton. If thoughts can travel that far, surely they can travel 200,000 miles to a space capsule, especially with no wind resistance.

The researchers comment that the sense of mental-physical causation might explain why people have a hard time pulling themselves away from the TV during critical moments of games. 'The experience of everyday magical powers makes people wary of cutting off their support at such key times, as they put off trips to the fridge and even avoid bathroom breaks in the pursuit of their team's success', they wrote. Your team *needs* you.

Sports fandom is a great way to show the everyday nature of magical beliefs, but in the most remarkable experiment by Pronin and her colleagues, the scientists revealed unsuspecting subjects to be would-be witch doctors.

A subject and a man playing another subject (a *confederate* in psychology lingo) arrive at the lab and are told they'll be exploring voodoo hexes. They draw pieces of paper to determine who will be the victim. The confederate pretends to draw the victim card, puts his name on a voodoo doll, and leaves the room. The subject focuses her attention on the victim and sticks five pins in the doll. When the victim returns he complains of a headache.

There were two conditions. In the neutral condition, the confederate acts naturally when he first arrives to the lab. In the 'evil thoughts' condition, he arrives ten minutes late, chews gum loudly, tosses his crumpled consent form on the floor, and generally behaves like an ass, arousing evil thoughts in the subjects. The icing on the cake: he wears a STUPID PEOPLE SHOULDN'T BREED T-shirt. By they time the subjects stick those pins in, they really mean it.

When asked whether they felt they'd caused the victim's headache, the neutral group, on average, said yes, slightly. We could mark that up to belief in mind over matter, or belief in the law of similarity, as the doll represents a person. But subjects felt much more responsible in the 'evil thoughts' condition, indicating that the more one's thoughts and intentions line up with a physical effect, the more credit or blame one takes.

So what explains such weird behaviour? According to the

researchers, apparent mental causation works the same way as apparent physical causation. Recall from the last chapter that we judge A causes B if A happens before B, A is consistent with B, and there are no other obvious causes of B. The trick is that this causation heuristic is so general that we don't distinguish between physical and mental causes. Event A could be an action or just a thought. If you picture something and then it happens, it's hard to fully dismiss your premeditation as an influence.

Such inferences are automatic and happen before you can even say hocus-pocus. But knowing that you draw such conclusions so easily means you can combat them. The next time a negative fate befalls a friend after you picture it – whether it's envisioned through worry or momentary ill will – you can tell yourself it wasn't your fault. You're just a needlessly guilt-ridden spectator.

Sky Mall

Mind over matter became a best-selling idea with the worst-kept secret in your airport book shop, *The Secret*. An Australian television producer named Rhonda Byrne made the *law of attraction* a household phrase in 2007 by stirring together the wisdom of several dozen self-help instructors, mixing in equal doses ancient mysticism and New Age optimism, and spiking it with artificial quantum-physics flavouring. The resulting potion: chicken soup for the conjurer's soul.

The law of attraction dictates that like attracts like, and according to Byrne, thoughts 'magnetically attract all like things that are on the same frequency'. Picture what you want in life, and the universe will provide it for you. 'It is exactly like placing an order from a catalogue', she wrote. Except negative thoughts become manifest too, so if you spend too much time on the plane mocking items in the Sky Mall catalogue, don't be surprised to find a Yeti garden sculpture on your doorstep when you get home.

Millions of people were instantly drawn to *The Secret*, seeking success or wealth or just a way to take control over their lives. But many found it hard to get past *The Secret*'s hyperbolic claims. 'Remember that your thoughts are the primary cause of everything,' Byrne writes. She claims that after three days of telling herself she could see clearly, she no longer needed glasses, and that she thought her way to being thin without a change of diet. An investor quoted in the book says he visualizes parking spots right where he wants them and 95 percent of the time they're there. (And yet, considering that 'the things you can think into existence are unlimited', one wonders why they stopped at good eyesight and prime parking instead of X-ray vision, a hovercar, and, why not, world peace.)

The Secret can also be criticized for encouraging readers to blame victims. Byrne wrote that even 'events in history where masses of lives were lost' resulted when 'the frequency of their thoughts matched the frequency of the event'. And there's no such thing as collateral damage, she argues – it's negative thinking that puts you in the wrong place at the wrong time. So are we to conclude that the infants on the *Titanic* brought their fate upon themselves? I don't recommend giving this book to a rape victim. Presumably, 'You were asking for it' doesn't come across any better when thought frequencies are invoked.

Finally, the book takes liberties with physical explanations that draw cries of 'foul!' from people who know their way around a particle accelerator. The New Age movement renders mind-over-matter miracles more palatable for many people by introducing the lingo and concepts of quantum mechanics. Research suggests that people find even bad explanations for phenomena more satisfying if the explanations include extraneous scientific-sounding information.

The law of attraction purportedly relies on a core feature of quantum physics called *wave-particle duality*. If you measure the particle-like properties of an electron such as its position, it will behave like a particle, but if you measure a wavelike property such as frequency, it will behave like a wave. What you're measuring in each case is 'the same phenomenon; it's just described two different ways, either as a wave or as a particle', says Victor Stenger, a particle physicist and the author of several books on scepticism. But having two possible descriptions creates the appearance that what you consciously decide to measure determines the nature of what you're measuring. 'People jumped on that and said, "Aha! Consciousness controls reality",' Stenger says. 'And that became the mantra of the New Age, starting a good thirty years ago.' The quantum mysticism movement took off in the 1970s with the books *The Tao of Physics* and *The Dancing Wu-Li Masters*, picked up steam in the 1990s aided by Deepak Chopra's writings, and gained new life in the aughts with the movies *What the Bleep Do We Know!?* and *The Secret*. According to Stenger, 'It's all just a big scam.'

Perhaps even more offensive to the scientifically minded, the law of attraction (as with most magic) is not *falsifiable*. Core to the scientific method is the notion that hypotheses must be subject to experimental scrutiny. A claim may be correct, but if there's no way one could even think of to disprove it, it's not a scientific claim. When the law of attraction fails to give you what you want, supporters argue that maybe you didn't focus enough or your worries got in the way; the problem is you, not the law. Sure, perhaps wishes come true, but if every failed wish draws excuses of 'you just didn't wish right', how can one ever put the idea to a fair test? Untouchable ideas (including theism, string theory, parts of Freudianism, and many of our deepest convictions) are sometimes said to be 'not even wrong'; they might be wrong, but there's no possible piece of evidence that could show it. And since they're compatible with any observable result, they provide no predictive power. They're the porcelain hammers of science: pretty but useless.

Despite serious drawbacks, the law of attraction has a large following. To keep myself honest in my evaluation of the law's merits, I talked to my friend Little Woo, a self-described spiritual alchemist in Vancouver who teaches workshops on the law of attraction (or 'manifestation', as she calls it). 'My life was difficult for a long time', she said. 'As I started to apply the law, I saw everything change for me. I saw the

manifestation of opportunities, of people, of really amazing circumstances.' For two years she wasn't dating and told the universe not to
send her anyone, and then one day she felt ready, and a week later 'I
manifested my partner', whom she's been with for several years. On a
smaller scale, one time she really wanted a specific dance shoe, in a
specific style and colour. (Woo is also an enchanting dancer and the
inventor of a practice she's dubbed 'burlesque yoga'.) Instead of going
around to shoe stores, she placed an order – with the universe. The
next day in dance class a woman stood up and said she had these brand
new shoes in a size 7 and did anyone want them? 'They were exactly
what I was imagining, exactly. Within twenty-four hours of the intention', Woo says. 'I have a lot of twenty-four-hour stories.' Happy sur-
prises like that were once rare, 'but now I can almost depend on them.
That's a very big thing for your life to be one miracle after another.'
Who am I to tell Woo and her students to return to a miracle-lite
world?

ıııııııııııııı

Scientifically, the law of attraction actually does work – but not through
mind over matter or quantum mechanics, just good old positive expectations and self-fulfilling prophesies.

Optimism helps you to see openings for success in ambiguous
situations and primes you to recognize opportunities when they come
your way. Hopes and expectations influence how we perceive the world,
even modulating the way the primary visual cortex processes raw
information. So they can make you see the glass as slightly more than
half full, both metaphorically and literally. And they push you to persist.

The psychologist Richard Wiseman has asked people who consider themselves either lucky or unlucky to look at a puzzle and decide
if it's solvable. Sixty percent of the unlucky subjects called it impossible, but only 30 percent of the lucky ones did. And if you'll recall
Lysann Damisch's research from the last chapter, people who felt luck

was on their side performed better while putting golf balls or forming words from a string of letters. Generally speaking, when you're confident in success, you see obstacles as mere speed bumps – or even opportunities in disguise. Believe in your ability to manifest the life you want, and you won't take no for an answer.

Experience conforms to expectations in the social world too. Expecting kindness from strangers will lead you to approach them more warmly, which then draws the kindness out of them. Or if you picture yourself getting a particular job, you'll walk into the interview with confidence and make a strong impression on the employer, which might get you that job. Little Woo may have been turning away potential suitors for two years through subtly off-putting body language. As soon as she opened her mind to the potential of falling for someone, her behaviour invited home delivery.

Finally, in the realm of health, hopefulness has been shown to reduce the severity of several illnesses, including cancer and heart disease, and optimism both boosts the immune system and leads to more health-enhancing behaviour: optimists exercise more, stick to diets, and follow treatment regimens.

To be clear, these benefits rely on good spirits and visualization, which is to say positive thinking, not magical thinking. Is there a benefit to going a step further and believing in the law of attraction as proselytized – to believe that positive thoughts have positive effects unmediated by behavioural changes?

Perhaps not. Rhonda Byrne decided she could lose weight without changing her habits and reportedly succeeded, but I suspect this result would elude replication for most dieters. And Little Woo's strategy for obtaining a pair of shoes omitted hitting the pavement and shopping for them. I posit she got lucky with her result. So it would seem letting your thoughts do the walking encourages one to sit at home and wait for the shipment. Not a reliable routine for success.

But. Let's look back at the research on luck. One might expect that feeling lucky would encourage one to passively allow good fortune to do its thing, yet this was not the case: people who feel lucky are motivated to actively capitalize on their gift, to strike while the iron is hot. So it's not unreasonable to suppose that faith in the magical power of one's expectations will lead one to run out and greet one's rewards in the field rather than hope they come to you. Even if you believe you can manifest a mate, you'll probably also recognize that a club or a book group sets a better stage than your basement den for such an encounter.

Further, believing that positive thoughts work wonders will lead you to have more of them, bringing all those proven benefits of optimism and positive visualization.

And, 'If you only activate thoughts, which is the visualization', Woo says, 'and you don't activate the emotional component – the feeling of certainty and trust – manifestation is a lot less powerful.' You need to believe in it for it to work, Woo is saying, but this remains true even if the law works through merely terrestrial means. It's the placebo effect. If you tell yourself to picture getting a job but don't have faith that the mental exercise will do anything, you'll walk into the interview feeling little different than if you'd done nothing. But thinking *I pictured it, and that means I'll get it* adds extra oomph. Again, positive thinking is the positive side effect of magical thinking.

We don't have nearly enough evidence on the effects of believing in mind over matter to draw sweeping conclusions about whether it's healthy or unhealthy. But I feel comfortable recommending optimism, as well as awareness that it can pay off in unexpected ways. If the results sometimes feel like magic, go with it.

Free Willy

There is one form of mind over matter that is completely accepted by nearly everyone even on an explicit level. In fact, to note that this phenomenon is illusory, as I have been known to do at cocktail parties, is to fight people's most basic intuitions about personhood, as well as to commit yourself to a long, sobering philosophical debate when you should be refilling your glass. The form of psychokinesis I'm referring to is the volitional control of your own behaviour. Free will.

I'm not the first to argue that the notion of bending your finger through force of will is no less magical than bending a spoon across the room. 'Is there any principle in all nature more mysterious than the union of soul with body, by which a supposed spiritual substance acquires such an influence over a material one, that the most refined thought is able to actuate the grossest matter?' David Hume wrote in 1748. 'Were we empowered, by a secret wish, to remove mountains, or control the planets in their orbit; this extensive authority would not be more extraordinary, nor more beyond our comprehension.'

You may argue that mental control over the body makes more sense than mental control over distant objects because the mind is contained within the body. Thoughts affect neurons, which prod muscles, which cause movement. It's all one system. But how do thoughts affect neurons? Thoughts are nonmaterial and neurons material. What would allow such a leap?

Some of the most-cited research in the case against free will comes from Benjamin Libet. In 1983, he and collaborators published a study in which they used an EEG to measure the brain activity of subjects as the subjects sat in a chair and spontaneously lifted a hand or finger. It was known at the time that electrodes placed on the head will pick up a signal called a *readiness potential*, or RP, at least several hundred milliseconds before someone performs an action. Libet wanted to compare the timing of this RP to the conscious experience of deciding to make a move.

During the experiment, subjects watched a spot of light revolve around a circular screen, like a fast second-hand on a clock, and reported its position at the time they became aware of wanting to move. They typically recalled deciding to act about 150 milliseconds before the movement. But the RP occurred 350 milliseconds before *that*. The brain had initiated action at least a third of a second before people had any awareness of the decision. Consciousness was just along for the ride, thinking it was calling the shots.

More recent experiments have extended Libet's findings. Scientists who placed electrodes in people's brains found preparatory neural activity a second and a half before subjects' awareness of their decision to act. And another group, using an fMRI scanner, recorded cortical decisions to press a button that took place up to ten full seconds before a subject was conscious of his decision, and they could even detect beforehand whether the subject would use his right or his left hand. If they'd analysed the data in real time, they would have known what the subjects were going to do seconds before the subjects themselves knew.

(After university I wrote an outline for a sci-fi film incorporating the exploitation of readiness potentials. As the protagonist, I had amazing fighting skills that drew not from reflexes but from anticipation. Through some fuzzy quantum entanglement scheme I could consciously read my adversaries' neural activity before they could and would, say, put my arm up to block a punch before it was even thrown. The film was to be called *Godspeed*.)

Libet refused to interpret his own findings as conclusive evidence against free will. He held on to the possibility of some kind of conscious veto power that could halt or redirect an act in progress – so-called free won't. But no available evidence supports such a magical intervention.

To some degree we are comfortable thinking of ourselves as automata guided by unconscious forces. We don't ponder the flexion of every muscle in our legs with each step. We accept that our desires and decisions are influenced by genetics, by upbringing, by culture, and by subconscious cues around us. But we hold on to the idea that we can trump all of that if we wish. That there is a magical imp inside us untethered by physics or biology or social pressures. That there is a ghost in the machine. But why?

David Hume noted that causality can't be directly observed. We can only infer causality by seeing two events in series. If a cue ball hits an eight ball, and the eight ball begins to move, it's likely the cue ball caused the eight ball's movement, but no microscope will pick up on any causality when those balls collide. Hume argued that we assume our intentions cause our actions because we regularly experience our intentions just preceding our actions. But even if B always precedes C, and is consistent with C, we still can't rule out something else, A, causing both B and C.

And Libet's research makes an A very likely. It appears that unconscious brain processes (A) build up to the conscious feeling of deciding to do something (B) and also actually doing something (C). A is unobservable to us, so when it causes both B and C we assume B caused C, fortifying our belief in free will. I thought and I acted. I am responsible. (This is a bit of a gloss. A conscious feeling of deciding can't influence the brain's judgement of whether it caused something. It's the neural activity underlying this conscious feeling that gets associated with the movement.)

'We are enchanted by the operation of our minds and bodies into believing that we are "uncaused causes", the origins of our own behavior', the psychologist Daniel Wegner of Harvard wrote in a book chapter titled 'Self Is Magic'. But knowing how 'The Great Selfini', as Wegner calls the self, works its most convincing trick doesn't dispel the illusion of free will. Every time you decide to lift a finger you're fooled again. Just as certain visual illusions still work even when you *know* the two lines are the same length, we still feel in control of our thoughts and actions. Even as I tell myself that I'm a passive observer of my own inner monologue, I'm fooled into thinking that I exercised free will in telling myself I'm a passive observer. So I correct myself, but, again, it feels like I freely chose to correct myself. The illusion is bulletproof. 'I'm a case in point', Wegner wrote. 'I've devoted years of my life to the study of conscious will. . . . If the illusion could be dispelled by explanation, I should be some kind of robot by now.'

Indeed, one could imagine a sophisticated robot doing much of what we do, with no conscious experience of free will. (In philosophy, such an imaginary being is called a *zombie*.) This robot would carry on a conversation by sensing vocal input, matching it to a dictionary, computing appropriate output, and synthesizing the corresponding speech sounds. This is much as the brain works (to a very rough approximation). So it would appear that consciousness is completely unnecessary, just an occasional side effect of advanced neural computation. (Computation more complex than any known computer can handle.) This view of the subjective mind as a mere by-product of brain processes with no actual power to do anything is called *epiphenomenalism*.

We typically see the interaction between mind and matter as a two-way street. The brain produces thoughts, and thoughts influence the brain. But with no evidence that thoughts can affect matter, and no theory as to how that would even be possible, we're left with the one-way street of epiphenomenalism. The brain produces thoughts, full stop.

But wait, doesn't the brain producing thoughts count as magic? Physical nervous systems produce subjective experience. So either we've found an exception to the definition of magic I set forth in the introduction or magic is real.

In a classic paper in the field of philosophy of mind, David Chalmers enumerates several challenges facing cognitive scientists and neuroscientists interested in consciousness. They must explain how we categorize stimuli, how we focus attention, how we report internal states, and so on. 'Getting the details right will probably take a century or two of difficult empirical work', he wrote in 1995. He called these tasks 'the easy problems of consciousness'.

The 'hard' problem of consciousness, he wrote, 'is the problem of *experience*. When we think and perceive, there is a whir of information-processing, but there is also a subjective aspect.' When we look at a red apple, our visual system registers the wavelength of light corresponding to what we call 'red', but there is also 'the felt quality of redness', he wrote. No one knows why felt qualities emerge from the detection of wavelengths. Or why the felt experience of weighing a decision emerges from the underlying neural computation of decision making. Many believe the hard problem is fundamentally unsolvable.

If anything is magic, consciousness is.

Autopilot

The *subjective experience* of free will may be robust (exempting special circumstances such as hypnosis), but we can more easily alter our *abstract conceptions* of free will, and those, believe it or not, have consequences outside Philosophy 101. A belief that the self is magic, that it freely breaks the chains of causality stretching back to the big bang, has implications ranging from personal responsibility to politics.

What happens when you tell people they're not magic? Kathleen Vohs and Jonathan Schooler invited subjects into the lab and had them reflect on a series of statements supporting determinism (e.g., 'A belief in free will contradicts the known fact that the universe is governed by lawful principles of science'), supporting free will (e.g., 'I am able to override the genetic and environmental factors that sometimes influence my behaviour'), or neither. Then subjects answered fifteen questions taken from graduate school entrance exams. They were asked to self-score and take $1 per right answer on their way out. Reading arguments for determinism led to a decreased belief in free will compared to the other two conditions and also led to taking more money. These subjects didn't actually perform any better than the others (the experimenters knew this from a comparison experiment), which means they were cheating – taking more money than they earned. 'To think that all behaviour is predetermined deflates the motivation to restrain one's urges', Vohs told me.

Studies by Roy Baumeister and others have shown that belief in free will reduces not only dishonesty but also selfishness and laziness. In one experiment, people who read statements supporting determinism said they'd be less willing to help people in several hypothetical scenarios (lending their phone to a classmate, giving money to a homeless person). In another study, workers at a day labour employment agency who had stronger beliefs in free will received higher ratings of workplace performance from supervisors.

'I think if you got people to not believe in free will it would probably also lead them not to exercise, and to drink too much alcohol', Vohs says. 'Maybe some people don't watch their weight because they think they have an obesity gene.' Being prosocial and proactive requires fighting base instincts, which requires the confidence that you *can* fight base instincts.

So as an epiphenomenalist I must be a useless jerk. And here I am trying to turn *you* into a useless jerk. (Can I interest you in my handsome line of DUALISTS SHOULDN'T BREED T-shirts?) But there are certain advantages to understanding the limits of willpower. For example, knowing the influence of certain cues over us allows us to avoid those cues. Alcoholics Anonymous teaches members to admit powerlessness over their addiction, and so people in recovery know to stay out of bars.

In a study showing the benefits of admitting powerlessness, researchers challenged participants to watch the Jim Jarmusch film *Coffee and Cigarettes* without lighting up. (This film is notable in part for a scene in which members of the Wu-Tang Clan discuss alternative medicine with Bill Murray.) First, subjects took a test and received bogus feedback labelling their level of impulse control as high or low. (Belief in one's ability to counter visceral impulses is not identical to a belief in free will, but they're related.) Then the subjects chose to subject themselves to one of four levels of temptation. If they didn't smoke during the film while a cigarette sat in another room, they'd earn two euros. They'd earn four for leaving one on a nearby desk, six for holding one in their hand, or eight for holding one in their mouth. Those who believed they had more control made the task harder for themselves, and in the end a third of them crumbled. The subjects who believed they had less control remained realistic in acquiring burdens, and with their easier loads only 11.5 percent lit up. Overestimating the power of will can make you underestimate the power of temptation.

Belief in free will has implications not just for inhibition but also for persuasion. John Bargh, the head of the Automaticity in Cognition, Motivation, and Evaluation (ACME) Lab at Yale, has spent much of his career exploring how subtle cues invisibly direct our behaviour. 'We're at the mercy of people who try to control us in ways we're not aware of', he says. Advertisers, for example. Or politicians. 'To the extent that we don't guard against those influences, we're just going to be pushovers.' He's shown that even an everyday action like holding a warm cup of coffee can affect us in surprising ways (in the case of the coffee, by making us behave more warmly). Millions of behavioural triggers surround us. 'I know everyone wants to believe these nice things about free will and agency', he says, 'and they do make us feel good, but they also can lead us into trouble.'

It seems the most reasonable strategy is to believe in a moderate amount of free will and to adjust one's estimations according to circumstances, such as levels of fatigue and social pressure. *Some* belief in free will is crucial for motivation and for resistance to temptation, but too much and you overlook hidden causes of behaviour, and think you can do anything. You're not a lump of clay, but you're not a god either.

Devil Inside

The danger of seeing people as supernatural extends to morality. Certainly, as a general rule, we should hold people responsible for their actions, but it's possible to hold people *too* responsible.

The law makes some accommodations for factors that reduce one's control over an outcome. Sentences are regularly reduced for damages caused involuntarily (manslaughter versus premeditated slaying) and crimes committed by those suffering mental illness or deficiency. And if I do something bad while under orders with a gun to my head, I'll likely go free.

But in most legal cases we grant people full blame for their actions. 'My brain made me do it' rarely works at trial. Neither can a perp pass the buck to a clockwork universe that sent him down his dark path. But *should* determinism enter the discussion? I submit that it should.

There are two main motivations for punishment. The first is retribution, commonly glossed as 'an eye for an eye'. *That bloke did me wrong, and I'm going to cause him harm in return*. It's an emotional and, in a way, barbaric response. The second type of motivation is utilitarian. Instead of looking back at what's been done, it looks forward: how can we prevent future offenses? Utilitarian punishments focus on deterrence rather than on deservingness.

Which philosophy of justice you prefer relies (or should rely) on your view of free will. If people have free will and are uncaused causes, then each of us is a *black box* – an entity with mysterious innards – and into this black box observers can impute mystical attributes of moral goodness or wickedness. If you believe in free will, then you should support retributive punishment. But if free will doesn't exist, then strictly speaking we're machines. And when a machine malfunctions, you don't call it evil; you fix it or set it aside. That's a utilitarian take.

Granted, terrible deeds throughout history have been justified by the framing of victims as mere objects, but thinking of humans as machines may in some cases promote compassion, as long as we allow that these machines can suffer. A passage from the sci-fi novel *Ender's Game* has stuck with me since I read it as a teen: 'In the moment when I truly understand my enemy, understand him well enough to defeat him, then in that very moment I also love him. I think it's impossible to really understand somebody, what they want, what they believe, and not love them the way they love themselves.' To understand is to forgive. And with enough patience, you can understand a machine, at least in theory. But you can never understand what's inside the black box of a freely acting Selfini. You can only praise it for its virtue or punish it for its vice.

So which model of punishment does our criminal justice system rely on? The primary currency of recompense is jail time, which serves both retributive and utilitarian functions – the offender suffers for what he's done, and he's also kept off the streets and used as an example for other would-be baddies – so society's possible motivations for imprisonment would appear hard to tease apart. A set of laboratory studies, however, has shown that while people profess to support both kinds of reasons for criminal punishment, when deciding appropriate prison sentences bloodlust reigns supreme. Apparently, we adhere to Immanuel Kant's prescription that criminals receive discipline 'proportionate to their internal wickedness'.

Could telling people free will doesn't exist actually change their moral judgements? The philosophers Shaun Nichols and Joshua Knobe asked people to imagine a universe 'in which everything that happens is completely caused by whatever happened before it'. For example, when someone named John ordered french fries, 'it *had to happen* that John would decide to have french fries'. When asked whether full moral responsibility was possible in such a universe, only 14 percent of respondents said yes. While nearly everyone agreed that this deterministic world did not resemble our own, convincing some of them otherwise might lead to a more utilitarian take on punishment. The trick would be to encourage logical deliberation and the suppression of emotional knee jerking. When Nichols and Knobe presented a lurid scenario in which a man burns down his house with his wife and children inside so he can run off with his secretary, 72 percent claimed that, even in a prix fixe universe, the man bears full moral responsibility. Presumably, they would also have liked nothing more than to see him choke on a chip for his efforts.

Our eye-for-an-eye instinct appears to be hardwired and likely aided the evolution of cooperative societies, where individuals know that defection has consequences. It seems a useful heuristic approximating civil rules of deterrence. 'If retributivism runs that deep and is that useful, one might wonder whether we have any serious hope of, or reason for, getting rid of it', the psychologists Joshua Greene and Jonathan Cohen wrote in a paper on neuroscience and the law. 'Have we any real choice but to see one another as free agents who deserve to be rewarded and punished for our past behaviours?' Indeed, a life without the heroism and tragedy of moral champions and villains is unthinkable.

But Greene and Cohen compare the belief in free will to the notion (debunked by Einstein) that space is flat. A flat picture of the world serves you well navigating the aisles of your supermarket but lets you down when calculating the trajectory of a satellite. Similarly, free will and retributivism make sense in many everyday encounters, but when contemplating something like the death penalty, it's worth pausing a moment and pulling out the slide rule.

Psychiatry has already made advances in this direction, according to David Eagleman, a neuroscientist and the director of Baylor College of Medicine's Initiative on Neuroscience and Law. 'We used to punish people and say if you're schizophrenic or depressed it's because you don't have enough free will or you're not being tough enough, and we're going to beat you', he told me. 'And over the last hundred years we've evolved to say you are your brain and there are organic disorders of the brain, and we now try to figure out how to fix those instead of punishing the person.'

According to Knobe, 'even if the automatic tendency to blame people never goes away, we could at least inhibit it to some degree in light of our beliefs about what's really right.' Deciding how to reform the system is a nontrivial challenge – rehabilitation might receive more emphasis, terms of incarceration might go up or down for various crimes – but it's one worth engaging in if we want to build a criminal justice system that isn't based on superstition.

Message Received

Edgar Mitchell, the astronaut, says people at NASA didn't make 'a big deal' out of the possibility that global well wishes aided Apollo 13's recovery; it was just 'around-the-water-cooler-type talk'.

There was similar talk after Apollo 14. Before the mission, for which Mitchell served as the lunar module pilot, he arranged with friends to conduct an ESP experiment while in space. He attempted to mentally transmit a series of abstract symbols, and correspondents on the ground attempted to receive them. According to Mitchell and his collaborators, the experiment was a success. Word soon got out, and 'there were quite a number of engineers and scientists at the space centre who would come to my office furtively looking up and down the hall and say, "Can we talk about that?"' Mitchell says.

We started this chapter exploring the hunch that thoughts can go out into the world and do real work. If the skull is permeable to the self, many believe the permeability goes both ways – we can directly *receive* information without using our physical senses. ESP, or extra-sensory perception, is the flip side of psychokinesis (PK) and consists of three main types: precognition, or seeing the future; clairvoyance, or seeing distant objects or events; and telepathy, the direct transmission of thoughts between minds.

Scientific studies endorsing each of these phenomena exist, but most scientists consider them unsound, or at least inconclusive. In any case, people don't believe in PK and ESP because of journal articles. They believe in them because of experiences they've had, or stories told to them by friends, or gee-whiz accounts of woo-woo in the media. You focus on a series of traffic lights and they turn colour as if on cue, or you dream about an earthquake and awake to find one has taken place a continent away. A series of small successes or a single extraordinary tale can challenge the stubborn sceptic in nearly anyone.

What all examples of PK and ESP have in common is a reliance on coincidences – in particular, shared features between thoughts and events (or thoughts and thoughts in the case of telepathy) – and coincidences are the manna of magical thinking. They also feed the law of similarity (see chapter 2) and superstitious rituals (see chapter 3), where events align with symbols and actions, respectively. For an organism that survives and thrives on identifying and interpreting patterns in the world, coincidences demand explanation, and extraordinary coincidences demand extraordinary explanations. Calling them chance doesn't cut it. In our quest for causes, we often find them in ourselves: supernatural abilities.

And what interpretation could be more satisfying? Psychic powers are a tempting explanation for egotistical reasons. We would love to believe that we have hidden capacities. Superheroes seduce us with their magical powers, and even conventional (but false) wisdom tells us we use only 10 percent of our brains. If we could just tap into our full potential, we think, maybe we could will things into existence or read people's minds or see the future. Personally, I would settle for remembering people's names after hearing them only four times.

Our search for extraordinary explanations for extraordinary events shows that we don't fully appreciate what the mathematicians Persi Diaconis and Frederick Mosteller have called the *law of truly large numbers*. In brief, they wrote, 'with a large enough sample, any outrageous thing is likely to happen'. If a remarkable coincidence happens to just one in a million people each day, that's over twenty thousand people a year in the UK alone. 'When such events occur, they are often noted and recorded. If they happen to us or someone we know, it is hard to escape that spooky feeling.' (As with the rest of this book, when I speak of 'belief' in a magical phenomenon, you can take that to include 'half belief'. A spooky feeling counts.)

Coincidences draw our attention and stick in our memories. Non-coincidences happen all around us all the time, and we either don't notice them or don't remember them. You thought about your mum and she called right then. Wow! But how many times a day do you think about your mum? Probably a lot (if you're a *good* son or daughter). We also tend to count close calls as hits. If you think about your mum and she calls an hour later, or your gran calls or a friend's mum calls or someone calls and mentions your mum, you might make a note. Our associative, connection-making minds will find a link.

Further, we think about a lot of people, and a lot of people call, or e-mail, or cross our paths in other ways. Unless you specify ahead of time that you are looking for a particular coincidence (your mum calling), experiencing it should probably not be the big deal you make it out to be. There are myriad opportunities for coincidences in our everyday lives; some loose ends are bound to link up now and then.

Of course sometimes there really is something to a coincidence. But it's not magic. For example, people (even strangers) finishing each other's sentences is to be expected on occasion, as we tend to share similar brain circuitry and cultural influences. And most people believe that dreams foretell the future, but if a prescient dream displays any real faculty (and that's a big if), it's not so much a sixth sense as an eye for subtle trends in one's environment. And what about the day in 2005 when 110 people won second prize in the US lottery Powerball, instead of the usual 3 or 4 people? Did they all select the same number because of a massive mind meld? Nope, the number was printed inside a mass-produced fortune cookie.

Coincidences that may not surprise anyone else will get a rise out of you if you're a star player in the event. Ruma Falk orchestrated a series of coincidences among a classroom of undergraduates and found that students were more surprised by coincidences that happened to them than by identical coincidences that happened to their classmates. Falk ascribes the phenomenon to an egocentric bias; you're unique, and what happens to you is more important than what happens to anyone else. She wrote that in telling her colleagues about her findings, she heard more than once, 'but you should hear what happened to me. . . .'

So we notice odd phenomena that appear to happen more than chance should allow, and we need an explanation. When someone has a psychic experience, 'it has a huge emotional impact', says Chris French, the editor of the *Skeptic* magazine and head of the Anomalistic Psychology Research Unit at Goldsmiths, University of London. 'And when some boring old sceptic like me says, "Well maybe it was just a coincidence", it really doesn't seem to be an adequate explanation. Certainly, it doesn't feel adequate on the emotional level, but also even intellectually.'

There's one simple explanation that can take care of a lot of diverse experiences: the mind can cause and sense things on its own. But 'the skeptic asks that what seems to be a unified – although not terribly deep – explanation of a host of phenomena (i.e., that psychic powers exist) be replaced by a patchwork of explanations or no explanation at all', the American psychologist Thomas Gilovich has written. 'The skeptical perspective, therefore, may sometimes be rejected because it can appear to lack elegance.' We have this mass of anecdotal data seeming to support PK and ESP, and it's really so much effort to go through one by one and say, well this isn't so rare, and that's the result of certain cognitive biases, and for this other thing, who knows? Believing in magic makes it all so much easier. Most scientists abide by the principle of Occam's razor: given two explanations for a phenomenon, they'll prefer the simpler one. But Occam's razor is double-edged. For me, attributing eerie encounters to a host of known mechanisms seems the simplest solution, whereas others might choose the shortcut of supposing a new, paranormal route.

You Lookin' at Me?

If people think they can receive transmitted thoughts, the most convenient device for transmission is the eye. Cultures around the world foster a belief in the evil eye – the power to curse a victim through a simple gaze inflected with envy or spite. And the sense of being stared at is nearly universal, even among educated citizens. Bruce Hood of the University of Bristol asked 219 university students who'd taken courses in vision and perception to rate their agreement with this statement: 'People can tell when they are being watched even though they cannot see who is watching them.' Only 4 percent strongly disagreed.

Direct stares kindle our nerves, especially when coming from a crush, a boss, or a stranger standing too close on the tube. Or when we have a spot on our chin. Hood suspects we interpret this emotional arousal as a transfer of energy. And the confirmation bias supports our sense of gaze detection; we remember times we feel we're being looked at and are correct, and we forget times we're wrong. Further validating the belief in sensed stares is the obvious fact that if you whip your head around to see if someone across the room is really watching, your movement will draw his attention even if he hadn't noticed you before. Chill out. No one sees your spot.

Awesome!

For the universe to speak a language the mind can directly perceive (through ESP), and for it to understand and obey the language of the mind (through PK), the mind and the outside world must share some common understanding; they must be on the same wavelength, so to speak. And while sometimes we feel a trickle of communication between them – a small wish fulfilled, an eerily prescient dream – occasionally the floodgates open. Where once you were sipping at a fountain (or peeing in a stream), now you're dissolving into a vast sea, expanding outwards, becoming one with everything.

Once Edgar Mitchell had completed his duties on the moon, and his co-cadet Alan Shepard had finished hitting his lunar golf balls, the two rejoined Stuart Roosa in lunar orbit for the return trip to Earth. On the way back, Mitchell had little to do but play tourist and stare out the window.

The spacecraft rotated to spread the sun's heat evenly, so every two minutes, the sun, the moon, the Earth, and the stars all passed before him. 'As I looked at that, I suddenly realized from my studies in physics and astronomy that the molecules in my body and the molecules in the spacecraft and the molecules in my partners' bodies were created in an ancient generation of stars', Mitchell told me. 'That was really a Wow experience, and it was accompanied by an ecstasy. I recognized the unity of things and recognized that everything is interconnected.' In his book *The Way of the Explorer*, he wrote, 'I was overwhelmed with the sensation of physically and mentally extending out into the cosmos. The restraints and boundaries of flesh and bone fell away.'

Mitchell's Wow experience continued for three days as he looked out the window. He found it so 'puzzling and powerful', he told me, that he set out to understand it upon his return. 'As I travelled around the world and located mystics and holy men, I realized that in every culture somewhere in the literature and the lore is a description of this type of transformative experience. In the Greek it's called *metanoia*, a change of mind. In Zen Buddhism it's called *satori*, enlightenment. In the ancient Sanskrit it's called *samadhi*.'

And for a small cadre of scientists, it's studied under the rubric of *awe*. In 2003, Dacher Keltner and Jonathan Haidt published a theoretic paper in *Cognition and Emotion* titled 'Approaching Awe, a Moral, Spiritual, and Aesthetic Emotion'. They proposed that the prototypical state of awe contains two central features: a *sense of vastness* – in size, complexity, beauty, power – and the *need for accommodation*, or the disorienting inability to make sense of what's happening by accommodating it with existing concepts and categories. One doesn't require a 360-degree view of the universe to experience awe. 'Songs, symphonies, movies, plays, and paintings move people, and even change the way they look at the world', Keltner and Haidt wrote. 'The same can be true of . . . skyscrapers, cathedrals, stadiums, large dams, or even oddities, such as the world's largest ball of string.' I've seen pictures of that ball, and I can confirm its awesomeness. It even has its own gazebo, albeit one not particularly awe-inspiring as far as gazebos go.

Were Paul on the flight path to Damascus, Mitchell might have interpreted his ecstasy in theistic terms, but the God-fearing do not have an exclusive hold on transcendence. In *Civilization and Its Discontents*, Freud wrote of correspondence with his friend Romain Rolland, a French writer who had studied Eastern mysticism. Rolland had described to him 'a feeling which he would like to call a sensation of "eternity", a feeling as of something limitless, unbounded – as it were, "oceanic". This feeling, he adds, is a purely subjective fact, not an article of faith . . . One may, he thinks, rightly call oneself religious on the ground of this oceanic feeling alone, even if one rejects every belief and every illusion.'

No less than the über-antitheists Richard Dawkins and Christopher Hitchens have copped to taking a dip, or at least stumbling into a pool, before regaining their footing on materialist ground. In *Unweaving the Rainbow*, Dawkins wrote, 'The impulses to awe, reverence and wonder which led Blake to mysticism (and lesser figures to paranormal superstition . . .) are precisely those that lead others to science. Our interpretation is different but what excites us is the same.' First comes the awe, then the (attempted) accommodation. In *God Is Not Great*, Hitchens noted that one needn't place bets on the 'supernatural' to experience the 'numinous.' And in several interviews and debates before his death he said that we have a need for the 'numinous', the 'transcendent', even the 'ecstatic', and that he wouldn't trust anyone who didn't respond in such a way to music, love, poetry, and landscape.

A state of transcendence in itself might not be called magical thinking; rather than attributing particular mental and nonmental properties across boundaries, one loses sight of the boundary itself – the line dissolves. Ego loss does, however, provide a catalytic cauldron for magical thinking. You don't *have* to interpret your mystical experience as a brush with a universal essence, but it's often hard not to.

In a study by Michelle Shiota, along with Keltner and Amanda Mossman, subjects instructed to recall a recent encounter with beauty in nature said they felt the presence of something greater than themselves and felt connected with the world around them. Another group of subjects, after staring at a seven-metre tyrannosaurus skeleton for one minute, were more likely than people who stared at an empty hallway to describe themselves as part of a larger system. And another study by Belgian researchers found that eliciting awe in subjects led them to believe more strongly that life has a purpose and to feel more committed and connected to humanity – they endorsed statements such as, 'There is an order to the universe that transcends human thinking', and, 'I still have strong emotional ties with someone who has died.' Overall, these are not 'I've become one with the Lord'–type reactions, but it's a nibble at the supernatural.

Abraham Maslow referred to ecstatic moments as 'peak experiences', equally accessible to atheists and clergy. Instead of God fearers and God deniers, here's how he divided humanity:

> The two religions of mankind tend to be the peakers and the non-peakers, that is to say, those who have private, personal, transcendent, core-religious experiences easily and often and who accept them and make use of them, and, on the other hand, those who have never had them or who repress or suppress them and who, therefore, cannot make use of them for their personal therapy, personal growth, or personal fulfillment.

Like many other psychologists, Maslow emphasized the importance of these peak-oceanic-Wow experiences for healthy development. Sometimes we just need to get beyond ourselves. Mitchell notes that several of the Apollo pilots found postflight callings in art, poetry, and environmental protection. An analysis of the memoirs and oral histories of dozens of astronauts found that having gone into space increased their sense of unity with nature, their desire for world peace, and their belief in God. And Mitchell himself founded the Institute of Noetic Sciences, which conducts and sponsors research on consciousness (including the possibility of distant healing through quantum mechanics). He's also taken up meditation, tapping a daily dose of awe through om.

Whatever the cause of your Wow, 'It is my strong suspicion that even one such experience might be able to prevent suicide', Maslow wrote. 'These deductions from the nature of intense peak-experiences are given some support by general experience with LSD and psilocybin. Of course these preliminary reports also await confirmation.' I have a personal data point I could add.

First I should say I'm what Maslow would call a nonpeaker. Perhaps this is why I really took to LSD when I began dosing. It allowed for feelings many other people probably take for granted. Emotionally, it made me a late bloomer instead of a nonbloomer.

My most transformative experience took place one morning in March 1996, driving back to Boston, Massachusetts, from a rave in southwest Connecticut. The night had been filled with moments both hilarious and terrifying. Leaving the venue, I purchased a mix tape by a trance DJ and popped it in my tape deck. I set out onto an empty highway, headed into the sunrise, the soundtrack filling the car. Notably, I still had LSD and MDMA in my system. (For the record, a combined acid-ecstasy trip is known as a *candyflip*. A combined acid-ecstasy-car trip is known as *really fucking stupid*. See also: *tempting fate*.) When trying to describe this experience in person, I have resorted to a gesture that involves arcing my hands upwards and away from each other while splaying my fingers. It was a new dawn. On this morning, after nearly eighteen years of life, I learned to accept myself and to love myself. Suicide would never again be an option. The depression didn't disappear completely, but the self-acceptance has remained, going on a decade and a half.

Several factors set the stage for my peak. The music, the sunrise, the open road: check. My psychiatrist had also just bumped up the dosage on an antidepressant I'd found to work for me. And I was learning to accept the love of a girl. But the serotonergic wildfire the drugs sparked in my brain suddenly brought to light unfamiliar feelings of connection and empathy. Days later I was still buzzing, still accommodating, writing in my journal (after smoking a bit), 'everything is vibrations', and 'why do I feel love? why do I feel Gratitude? these are strange emotions for me. Why do I feel FREEDOM?' To accommodate the strange emotions – I was pretty sure I hadn't felt love towards anyone in a decade, let alone said 'I love you' – I had to restructure my views of myself and of reality (what physicists would call a phase change), and fortunately the new gestalt stuck.

Of course now, and even in the moment, I could parse the experience in terms of neuroscience and psychology, but understanding light refraction does not reduce the magical power of a rainbow after a long storm.

To Infinity and Beyond

An expansive concept of the mind incorporating psychokinesis and ESP – while generally unfounded and often counterproductive – can proffer feelings of both control and meaning. Believing that our thoughts have the power to drive our own behaviour as well as the behaviour of the outside world – that they're not just feeble shadows cast against the inside of our heads – provides a sense of agency and makes us go out there and become active participants in life. And the notion that our thoughts leak out, or that we can tap into others' minds or detect events from afar, gives us the feeling that we're tuned in to the world around us, and that we can communicate and bond with people in profound ways.

Then there's the Wow factor, which one can exploit without becoming a full-fledged mystic. Oceanic feelings inspire us to write poetry, fight for world peace, and pursue scientific curiosity. They can even teach an emotional delinquent about real human connection. While mystical states may not unite you with a universal intelligence, they can still tap you into your own potential for transformation.

Which makes them mind-expanding after all.

5

The Soul Lives On

Death Is Not the End of Us

One day in school a boy from the year above mine approached me and handed me two dollars.

'What's this for?'

'Just take it.'

'Why?'

'I can't tell you. Just take it.'

Later that day a friend of his came up to me, and we repeated the exchange. Then another. And another. I now had $8 for no apparent reason.

Eventually, I got the full story. These four boys had been playing with a Ouija board the night before and had evidently reached a spirit named Sarah, who revealed herself to be an ancestor of mine. Sarah, after spelling out my name, provided the reason for her call. She promised them a lifetime of good fortune if they each paid me the reasonable sum of $2 million. The guys, rather than hang up, decided to negotiate, eventually driving the price down to $20,000, then $20, then $2. Either Sarah didn't have a knack for driving a hard bargain (alas, evidence of family resemblance) or they could dismiss her existence with only 99.9999 percent confidence. Either way, her intervention did not make headlines, which cannot be said for that of Harry Fuller.

||||||||||||

On 10 February 1993, Harry and Nicola Fuller were found murdered in their home in East Sussex. The killer had shot Harry once in the heart through his back. Nicola was shot three times and left for dead. As she telephoned for help, speaking through a splintered jaw, the killer came back into the room, placed a duvet over her head, and finished her off.

Several lines of evidence pointed to a single suspect: a man named Stephen Young. He'd arranged to visit Fuller that day, his car was seen in the area, he had matching bullets in his home, he knew Fuller to keep large amounts of cash on hand, and he made a significant bank deposit the next day. It seemed an open-and-shut case of cold-blooded murder. After a three-and-a-half-week trial, the jury returned a unanimous guilty verdict on the second day of deliberation. The courtroom cheered.

But something happened between the two days of deliberation that threw a major kink in this case.

On 22 March 1994, the jury retired to a local hotel. Over dinner the topic of Ouija boards came up. At 11 P.M. the bailiffs escorted the jurors to their bedrooms, but four jurors reconvened and set up a séance for shits and giggles using a drinking glass and scraps of paper with letters written on them. They each put a finger on the glass and moved it around. Here is how one of the jurors described the proceedings in an affidavit:

> Ray then asked, 'Is anyone there?' The glass went to 'yes.' Ray said, 'Who is it?' The glass spelt out 'Harry Fuller.' . . . Ray said, 'Who killed you?' The glass spelt out 'Stephen Young done it.' Ray said, 'How?' The glass spelt 'shot.'. . . Ray also asked who killed Nicola, and the glass spelt out 'Stephen Young.'

There was additional discussion of the gun and the money.

We then discussed among ourselves what we should do and the glass spelt out 'Tell police.' I said 'We can't.' It then spelt out 'later, us and you.' It continued, 'Vote guilty tomorrow.'

During this time Ray made notes. It is only right to say I was crying by this time and the other ladies were upset as well. We realised it had gone too far and we ended the exercise.

These four jurors discussed their experience with the others the next morning at breakfast, and by the end of the day Stephen Young had received two life sentences.

The story would have ended there, known in full to only twelve people, but in early April one of the jurors who hadn't taken part in the séance reported the event to a lawyer, who contacted Young's lawyer, who then appealed the conviction. On 24 October, the Court of Appeal, Britain's second-highest court, ordered a retrial. The Lord Chief Justice filed this reasoning:

It seems to us that what matters is not whether the answers were truly from the deceased, but whether the jurors believed them to be so or whether they may have been influenced by the answers received during this exercise or experiment.

Was it merely a drunken game which the court should disregard, as Mr Lawson [on behalf of the Crown] suggests? We do not think it can be laughed off in that way.

In December, a second jury found Young guilty, this time based purely on terrestrial evidence.

||||||||||||||||

Across cultures and eras, afterlife beliefs are nearly universal. Whether you expect to waltz through heavenly gates, unite with a collective spirit, take the form of another creature, or spend your evenings toying with Ouija players, you likely believe The End is really just a To Be Continued.

But why are we in such denial of death? This question has haunted psychologists for ages.

Body and Soul

Scientific evidence for an immortal soul leaves something to be desired. By most accounts the mind is a product of the brain. (Alternately conceived, the mind *is* the brain, the mind is a *property* of the brain, or the mind is what the brain *does*.) Injury to the brain can cause cognitive deficits, drugs that pass the blood-brain barrier can affect one's mental state, and electrical stimulation of the cortex can produce hallucinations. In short, what happens to the brain happens to the mind. So it makes logical sense that when the brain stops, the mind does too.

But does it make intuitive sense? Some psychologists believe we are natural dualists – we instinctively dichotomize the world into two types of substances: thought and matter, body and soul, *mind* stuff and *stuff* stuff.

Dualism does have an obvious rightness about it. Mental life clearly escapes the constraints of the physical world: our thoughts and feelings and sensation remain private, they can change or disappear in an instant, they can't be handled or measured the way objects can, and they can represent things that don't actually exist.

We pick up on these unique qualities of mind early. Children at least as young as three can correctly explain that, unlike a real biscuit, an imagined biscuit can't be looked at, touched, or eaten, that it can't be shared with a friend, and that it can't be saved for another day. Further, they recognize that unlike red flowers and dogs that roll over, talking flowers and flying dogs can only be thought about and are not really real. Not only can toddlers distinguish between the mental and the physical; they can reason about and articulate the differences. They have metacognition.

Distinguishing internal from external phenomena – a useful skill to pick up early in life – might be enough to lead to dualism, but not necessarily the kind required for belief in an afterlife. Rather, it demonstrates no more than a weaker form of dualism called *property dualism:* the brain has both physical properties (mass, volume) and mental properties (consciousness), but if the brain is destroyed, all its properties go along with it.

Recognizing that a thing can make a thought – that your three-pound brain produces subjective experiences (the image of a flying dog, the desire to own such a dog, sadness that such a dog does not exist) – doesn't mean you believe the thought can fly off without the neurons keeping it aloft. Like a shadow and an obstruction, a thought can't be equated with the thing that casts it, but can't be separated from it either. Most people who call themselves materialists are technically property dualists, unless they deny their own consciousness.

The kind of dualism required for an immortal soul is called *substance dualism,* or *Cartesian dualism,* after René Descartes: the mind is a different type of substance from the brain and can exist independently of it. This is the kind of dualism people usually have in mind when they speak of dualism, and it's the kind meant by psychologists who argue that we're innate dualists. (See, for example, Paul Bloom's wonderful book *Descartes' Baby.*)

The strongest corroboration of the claim that we're natural dualists may be the ubiquity of afterlife beliefs. A soul can exist beyond brain death only if it's an independent substance. Yet one could argue that belief in an afterlife (and dualism) is not innate but merely learned, that it's an appealing and viral idea that humans everywhere cling to; because of their fear of death people invent the idea of the immortal soul and then teach it through religion. If that's true, most of us are substance dualists, but we're not *naturally* substance dualists. The psychologist Jesse Bering, of Queen's University, Belfast, in collaboration with David Bjorklund, decided to test this argument. How did they attempt to get to the bottom of our deepest doctrines on the nature of the spirit and the great beyond? By putting on a puppet show. That's how I would have gone, too.

Except instead of bashing Transformers together, Bering put some thought into his show. He presented a narrative to five-year-olds, eleven-year-olds, and nineteen-year-olds: 'One day, Baby Mouse decides to take a walk in the woods.' A finger puppet enters a scene decorated with fake grass and water and flowers. The subjects are told about what the mouse is thinking and feeling. Enter hand two. 'The bushes are moving! An alligator jumps out of the bushes and gobbles him all up. Baby Mouse is not alive anymore.' Just when we were getting to know him. The children and adults then answer several questions about the dead mouse's biological and mental functions.

The first thing Bering and Bjorklund found was that even when people realized the mouse's body had met its match, they didn't give up on his mind. A full 88 percent of five-year-olds acknowledged that Baby Mouse's brain no longer worked, yet 94 percent believed he still loved his mum, and 79 percent claimed he somehow knew he was dead. Among adults, 100 percent recognized the mouse would never again need to eat food or drink water, but 60 percent said the mouse still believes he's smarter than his brother. (Sibling rivalry dies hard.)

The second thing they noticed was that cultural indoctrination could not be the only reason for belief in the continuity of mental states after death. If it were, more adults than children should answer in accordance with afterlife beliefs. But the data revealed the opposite pattern. For example, while 77 percent of the five-year-olds said the alligator's cute furry lunch still hoped to get better at maths, the percentage went down to 55 for the older children, and 12 for the adults. Further, the experimenters note, very few children used religious terms such as *heaven*, *spirit*, or *God* during the study.

If we are natural Cartesian dualists, what might be the source of this mystical belief? As I mentioned above, the recognition of internal states as different from external ones (imagined biscuit versus real biscuit) might get us there, but it only guarantees us a ticket to dualism lite, where mind is separate from body but still dependent on it. I believe a stronger impetus for full-fledged dualism is the illusion of free will.

If you believe you have an ego piloting your brain, then your mind is not simply a property of your body or a product of its neural activity but an independent player, an uncaused cause. Your soul makes its own decisions and then tells the body what to do. It's its own man, unbeholden to the workings of biology. And if it can *act* independently of the brain, then it should be able to step out of the cockpit and *exist* independently of the brain. Which means the mind is free to do whatever it wants – and *go* wherever it wants – once it's done inhabiting its mortal vehicle. In a brief article attributing belief in an immortal soul to the illusion of free will, three psychologists punctuate their commentary with this kicker: 'James Brown, move over: Conscious will is the Godfather of soul.'

Recall from the last chapter that the illusion of free will rests on the principles of priority, consistency, and exclusivity we use to judge causality. In the simplest terms, we believe in free will because we experience thoughts immediately before we experience associated actions and thus judge our thoughts to be responsible for the actions. This means that a lush bouquet of myths and gospels depicting the journeys of the soul once released from its terrestrial duties flourishes out of the simple and generic rule of thumb that A causes B if A happens before B.

Understanding how people think about dualism might seem of importance only to religious scholars and academic psychologists. (Oh, and screenwriters, whom we can thank for everything from *Ghost* to *Ghostbusters*, *The Sixth Sense* to *The Seventh Seal*, *Freaky Friday* to *Friday the 13th*.) But doctors, politicians, and lawyers have reason to attend airy discussions about the soul; practical and ethical matters of real substance are at stake. Every decision we make regarding the creation and destruction of human life ends up on the table, most heavily those regarding abortion, capital punishment, and euthanasia. And between the bookends of birth and death we must handle notions of free will and personal responsibility, of brain damage and mental illness, of animal cruelty, designer pharmaceuticals, and (someday) artificial intelligence. Understanding the comings and goings and interventions of souls – or, more important (and more feasible), understanding our *conceptions* of the comings and goings and interventions of souls – is ultimately of vital importance to a functioning body politic.

The Impossible Certainty

For a period of my young life, I used to call for assistance when a nightmare woke me. I would yell for Mom or Dad until they came to my bedroom to comfort me. One night, when I was maybe seven, I asked my dad if dead people had dreams. I pictured death as sleep, and my biggest fear was an eternity of nightmares with no hope of rescue.

Try to picture what will happen when you die – accepting for the moment that there's no afterlife. Maybe you imagine lying in a coffin. Maybe you envision floating in a black starless void. Maybe it's a time of contemplation, reviewing your life in playback. However you picture the abyss, you're still picturing *something*, and that's not really an abyss.

Freud wrote in 1915 that it's impossible to imagine our own death: 'Whenever we attempt to do so we can perceive that we are in fact still present as spectators. Hence . . . at bottom no one believes in his own death.' Around the same time, the Spanish writer Miguel de Unamuno wrote: 'Try, reader, to imagine to yourself, when you are wide awake, the condition of your soul when you are in a deep sleep; try to fill your consciousness with the representation of no-consciousness, and you will see the impossibility of it. The effort to comprehend it causes the most tormented dizziness.'

For generations, the difficulty of mentally simulating one's own absence has been offered as a reason for our belief that such an absence is not possible, that something must come after the moment of death. As a PhD student, Jesse Bering decided to test this notion that constraints on our imaginations encourage afterlife beliefs, which he refers to as the *simulation constraint hypothesis*.

Bering presented a story to a group of university students, this time without the aid of puppets. A history teacher named Richard Waverly was driving to work. The vignette described how Richard felt and what he thought – he was tired, he was angry at his wife, he was worried about his lesson, and so on. In a moment of distraction, Richard hit a utility pole and died instantly. After reading the story, students answered six types of questions about the state of the deceased: biological (e.g., Will he ever need to eat again?), psychobiological (Is he still hungry?), perceptual (Can he hear the paramedics?), emotional (Is he still angry at his wife?), motivational (Does he want to be alive?), and epistemic (Does he still believe his wife loved him?).

Bering argued that the first three types of questions – biological, psychobiological, and perceptual – get at experiences we can easily imagine being without. We've all been in situations where we weren't hungry or where we couldn't see anything. If you haven't, just eat a sandwich and close your eyes. But emotions, desires, and beliefs are harder to escape and thus harder to simulate the absence of. Bering found that subjects were much more willing to attribute these latter states to the dead Richard than the first group of states. A majority allowed him the ability to be happy, to miss his wife, and to know he's dead. Further, even when correctly answering that Richard lacked a particular state, subjects took longer to deny this second group.

When indicating their explicit afterlife beliefs, 13 percent of Bering's subjects ticked the box next to this statement: 'What we think of as the "soul", or conscious personality of a person, ceases permanently when the body dies.' But even these 'extinctivists' had trouble letting go of Richard. Thirty-two percent of their answers regarding emotions allowed for the continuance of those emotions after lights-out. Same for desires. And 36 percent allowed for continued epistemic states. Bering wrote in his entertaining book *The Belief Instinct:* 'One particularly vehement extinctivist thought the whole line of questioning silly and seemed to regard me as a numbskull for even asking. But just as well, he proceeded to point out that of course Richard knows he is dead, because there's no afterlife and Richard sees that now.'

The study certainly supports the simulation constraint hypothesis, but, as Bering acknowledges, the data also lend themselves to another interpretation. Mental processes such as hunger and sensory perception have obvious connections to the body. If you die and your eyes and stomach stop working, of course you won't see paramedics or feel tummy grumbles. But higher cognition relies more exclusively on the brain, a more mysterious organ. Several subjects told Bering that psychobiological and perceptual states were 'physical things' but that emotions, desires, and beliefs were 'spiritual things'. Bering's experiment doesn't resolve whether some things are more 'spiritual' because they're harder to imagine being without or whether they're harder to imagine being without because they're more 'spiritual'. Do we picture consciousness without a body because we can't picture unconsciousness or just because the destruction of the body gives us no reason to picture the destruction of consciousness (thanks to dualism)?

Aside from difficulties proving the simulation constraint hypothesis, the idea also has some theoretical complications. If we believe we'll always exist in the future because we can't imagine not existing, shouldn't we also believe we always existed in the past? Yet afterlife beliefs far outnumber before-life beliefs in the world's cultures.

The philosopher Shaun Nichols suggests that the difficulty imagining nonexistence might work in tandem with a bias that focuses our attention on the future, and this might explain our lack of prelife beliefs. He cites the philosopher Thomas Nagel, who argues that even though we can't imagine nonexistence before birth, we don't really care if we didn't exist then. Nagel, in turn, cites the philosopher Derek Parfit, who has written about our asymmetric attitudes towards the past and the future. For example, most people would much rather have ten hours of painful surgery in their past than one hour of painful surgery in their future. As Nagel notes, facing the fact that we weren't alive two hundred years ago doesn't bother us as much as the prospect of our own elimination.

There's another potential theoretical objection to the simulation constraint hypothesis. The hypothesis is that we believe in the afterlife because we can't simulate a lack of thought. But we can conceptualize a lack of thought, in abstract terms. So if we can *know* we won't think or feel anything after death, why does it matter that we can't use our imagination to *simulate* it?

I don't know that this can be answered, except to say that it does matter. Nichols addresses the gap between conceptualizing and simulating nonexistence, writing, 'we see death all around us, and frankly, it's a little disconcerting. I can recognize glancingly that this suggests that at some point I *won't* exist. But . . . when I try to imagine I *don't exist*, I return to the dizzying imaginative obstacle'.

Everyone runs aground on this barrier. Gail McCabe, the humanist officiant quoted in chapter 2, described to me an exchange at a humanist discussion group: 'People got to talking about their funerals. I remember one fellow saying, "Well, my children better follow my plans for my funeral." I said "Wait a minute, you don't believe in an afterlife, why do you give a damn? You're not going to be there!"' Caring about enjoyment of your funeral is magical thinking. Once you die, nothing can matter to you. Not even the fact that you're dead. There will be no you for anything to matter to.

So if we can't escape this form of magical thinking and convince ourselves consciousness really ends, why are we still scared of death? Well, even if you can't picture the end of life, you can picture a radically transformed way of life. The other side might be a land of virgins and candy floss and family on their best behaviour, or it might be an eternity of awkward blind dates and liquorice and being about to sneeze. You'll be cut off from people and places and activities you love here on Earth, as well as whatever else you hoped to accomplish with your life. You will not fulfil all your potential. Make no mistake: no one will bother completing that ice-lolly-stick model of Big Ben you considered your legacy.

So death may be impossible to grasp fully, but what we know about it – and what we don't know about it – can still scare the hell out of us.

Knock, Knock

In 1797 Lord Horatio Nelson, one of the most decorated officers of the Royal Navy, lost his arm in the Battle of Santa Cruz de Tenerife. A musket ball passed through bone and tissue above the right elbow, and he immediately ordered the damaged arm to be removed, before resuming command thirty minutes later. At some point thereafter, Nelson began to feel his fingers painfully digging into his absent palm. He concluded that the body is not necessary for awareness to continue, calling his mysterious pain 'direct proof of the existence of the soul'.

In 1866, the *Atlantic Monthly* magazine ran a first-person article titled 'The Case of George Dedlow'. Dedlow, a doctor and a soldier, describes the loss of his arm in the American Civil War. A musket ball passed through bone and tissue below the right shoulder, and he had the arm removed, before returning to battle thirty days later. Then in the Battle of Chickamauga, on the Georgia-Tennessee border, he injured both legs. After surgery he asked someone to massage a cramp in his calf, only to learn that his legs were now gone. Eventually, he lost his left arm to gangrene, only to feel intense pain in his left pinky. 'Often, at night, I would try with one lost hand to grope for the other', he wrote, or rather dictated.

While in recovery, another patient takes him to a séance, where he's asked to think of a spirit for the medium to summon. Dedlow, a sceptic, gets a 'crazy idea'.

The medium asks if a spirit is present and hears two knocks – they have two visitors. She pulls out a miniature Ouija board, a card covered with letters and numbers. She points at each one until she hears a knock and transcribes the character. Finally, the phantoms identify themselves: 'UNITED STATES ARMY MEDICAL MUSEUM, Nos. 3486, 3487'. The medium looks confused. 'Good gracious!' Dedlow exclaims, now less sceptical, 'they are *my legs – my legs!*'

Upon publication of the story, donations and visitors began arriving at the hospital that accommodated Dedlow in Philadelphia. The only problem: he didn't exist. Readers had bought into the tale, despite the fact that soon after hearing the knocking Dedlow claims to have gotten up and walked across the room on his returned but invisible legs – a bit drunkenly for they had spent the previous nine months stored in alcohol.

The first modern medical account of 'phantom limbs', and the coining of the term, came six years later in a medical text by Silas Weir Mitchell. Mitchell specialized in damaged nerves, and in his treatment of Civil War soldiers he found that 'nearly every man who loses a limb carries about with him a constant or inconstant phantom of the missing member, a sensory ghost of that much of himself'. Mitchell, it had been revealed at this point, was George Dedlow's ghostwriter.

While phantom limbs themselves touch in a ghoulish way on dualism, I summon them here to help explain another type of phantom. We now know that even after a body part is gone, it's still represented in the brain; the area of the somatosensory cortex dedicated to receiving input from the part still exists and can be activated by stray neural signals. In the same way, when a loved one dies, you maintain a representation of that person's mind that defies extinction. For days or months or years after the death, you may continue to imagine what that person is doing right now, or wonder how he'll react to your latest news, before stopping yourself.

In *The Year of Magical Thinking*, Joan Didion describes her months of mourning after her husband John Dunne's sudden death in 2003. At one point, she is collecting his clothes to give away:

> I stopped at the door to the room.
> I could not give away the rest of his shoes.
> I stood there for a moment, then realized why: he would need shoes if he was to return.

Didion does not retain the shoes for her phantom legs. She retains them for her phantom husband. 'The recognition of this thought by no means eradicated the thought', she wrote.

The anthropologist Pascal Boyer has written that each of us

maintains 'a *person-file system*, a kind of mental Rolodex or *Who's Who* of [our] social environment', and even when a real person is deleted from life, the person's file remains in our heads. 'It keeps producing inferences about the particular person on the basis of information about past interaction with that person, as if the person were still around. A symptom of this incoherence is the hackneyed phrase we have all heard or used at funerals: "He would have wanted it this way." '

Jesse Bering has argued that 'offline social reasoning' – thinking about others without them present – contributes to our belief in the immortal soul. When people are not around, we don't assume they have ceased to exist – we don't delete their file. Even infants expect to find a person (or a toy) still sitting behind a screen that has temporarily blocked their view. We expect absent people to be out there doing something, living their lives. And what others think matters to us, so we frequently simulate their minds within our own. I often carry on imaginary conversations with friends or family members or acquaintances, playing the role of both parties myself. Our brains are populated with the disembodied souls of myriad other people.

When the bodies of those people fail, their minds remain intact in our heads. We receive no continued input from them and may even witness their cold flesh placed underground – just as phantom-limb sufferers receive no sensation from the limb and can clearly see that it's gone – but we still hear the departed speaking to us, carrying on their end of the conversation.

'Out-of-sight, out-of-mind is not the usual behavior of humans,' Mitch Hodge, a philosopher who studies afterlife beliefs, wrote in a paper. He goes on:

> As personal anecdotal evidence of this effect, shortly after the death of my father, I felt an overwhelming compulsion to call him because I had significant news. It was part of our normal interaction to call one another when there was significant news to report. This time, however, the news that I wished to relay to my father was that he was dead and that I would be flying home to attend his funeral!

Just as stray activity in amputees' brains is projected outwards to form a ghostly limb, neural activity in the grieving can project outwards to place a missing loved one back in the world, or in the netherworld, heaven, hell, or wherever he can continue being the person you remember. Boyer writes, 'We should not be surprised that the souls of the dead or their "shadows" or "presence" are the most widespread kind of supernatural agent the world over.'

It's unclear whether offline social reasoning on its own would be enough to convince anyone a lost loved one is not completely lost, but it certainly guides how we conceptualize the afterlife – a realm where inhabitants retain their personalities and their human form, perhaps with an upgrade of wings.

And it primes us to see messages from the beyond. In *American Scientist* magazine, Bering describes sitting with his siblings in his mother's living room the morning after she died: 'Just then, the wind chimes outside my mother's window started to sound. We looked at one another, and I, the family skeptic, knew exactly what was going through everyone's heads: "That's her! She's telling us not to worry!" I knew because I was thinking precisely the same thing.'

Thirty-four percent of Britons report believing in ghosts, and 37 percent say that 'restless spirits' exist, according to a Populus survey. Forty-three percent of the respondents also report that they have communicated with the dead.

Most ghost stories aren't cinema-worthy, let alone enough to compel Bill Murray to dust off his proton pack. They usually amount to hearing a wind chime or footsteps, seeing a shadow move in the corner of one's eye, or feeling an eerie presence. Such encounters implicate nothing more suspect than draughts, knocking radiators, or pets, plus anxiety, expectation, visual illusion, selective memory, and imagination. Some ghost-myth busters have offered brain-altering magnetic fields or chill-inducing subsonic frequencies as additional explanations.

In extreme cases, there's also the catastrophic combination of conmen and lonely prey. Elisabeth Kübler-Ross, the psychiatrist who earned notoriety by introducing the (now partially debunked) five

stages of grief in her 1969 book *On Death and Dying*, established a
spiritual retreat near San Diego in the 1970s, partnering with a supposed psychic named Jay Barham. During séances, visiting entities
sometimes had sex with female participants, who remained trusting
despite the entities' occasionally burping or mispronouncing the same
words Barham did. 'I needed to believe', a victim said later. Several
received vaginal infections. Eventually, one woman pulled the tape off
the light switch during a séance and found Barham wearing nothing
but a turban. Not all horror stories involve real ghosts.

Limited Previews

Just as with other paranormal phenomena, belief in the afterlife can
result from personal experiences or persuasive accounts from others.
Dramatic chronicles of reincarnation, possession, or haunted houses
go a long way.

As I type this, a search on Amazon for 'near-death experience'
brings up about a thousand books, nearly all taking NDEs to be evidence
of life after death. Between 6 and 12 percent of people who survive cardiac arrest have an NDE. Proponents of the supernatural interpretation
argue for the veridicality of these experiences on several points. They're
nearly universal across cultures, they feel at least as real as reality, and
people report witnessing events that took place around them while they
were clinically dead.

The psychologist Susan Blackmore addresses these claims in her
book *Dying to Live*. For example, many of the common perceptual and
memory features (the dark tunnel, the flashbacks) may result from distinct activation or deactivation of various structures in the brain such
as the visual cortex and temporal lobe. The sense of realism could
result from the mind settling on the imagination as real because external perception has become unreliable. And memories of events taking

place in one's vicinity while the brain is inactive could be reconstructed recollections based on guesswork.

NDEs and similar experiences can result from cardiac arrest, anaesthesia, head trauma, psychedelics, intense fear, or sometimes no apparent cause at all. The neurologist Kevin Nelson argues that NDEs involve the intrusion of rapid eye movement (REM) sleep into waking consciousness. Surely, you've had these half dreams, where you start to nod off during a lecture and the speaker grows tentacles.

Out-of-body experiences (a signature element of NDEs but not restricted to NDEs – they occur to about one in twenty people) can be convincing demonstrations of mind-body dualism. How could you be conscious and yet stare at your own body from an outside perspective if your soul weren't free to roam? But scientists have offered physiological explanations for OBEs too. The Swiss neurologist Olaf Blanke argues that disruptions in the functioning of the brain's temporoparietal junction cause conflicts between touch, vision, balance, and sense of body position. Confusion can also happen when input from one of these senses becomes weak or conflicts with other senses; out-of-body-type experiences can be induced through relaxation techniques or virtual-reality experiments. One group of neurons says one thing, another group says another, and suddenly you're on the ceiling. You've got a phantom self.

One way to test whether out-of-body experiences are real or just in the head is to place a message on a high shelf in a hospital room and ask resuscitated patients if they saw it. So far no one who's conducted such a study has reported positive results.

Undo-alism

To believe in the immortal soul is to believe in the supernatural, but is it magical thinking, as defined in the introduction? To the degree that we dualistically separate body and soul, it isn't. But that definition (*magical thinking is the attribution of mental properties to nonmental phenomena or vice versa*), while concise, isn't perfect, and I think ghosts and the afterlife bear enough thematic resemblance to the other forms of magical thinking to deserve inclusion. And in any case, we're *not* pure dualists; there are important ways in which we treat the soul as a substance and our meat as mystical.

Even the term *substance dualism* gives the game away. It's an own goal, as the word *substance* evokes materiality even when following the descriptor *nonmaterial*. We say the world consists of material and nonmaterial substances, but we often just think of two kinds of material substance, one wispier than the other.

Consider the way many people talk about the soul upon death. They say it *leaves* the body, and the *departed* is now *in a better place*. So it seems to always have a location in space, a defining property of physical substances. When we're alive, that location, we presume, is in the head. Descartes even picked a specific site; he decreed a small neural feature called the pineal gland the 'seat of the soul'.

Talk of *seats* and *better places* exposes more than linguistic convention. As we saw in the second chapter, metaphors reveal the very structure of our thoughts; even abstract concepts are grounded in our physical experiences. For example, we think of morality as purity, importance as weight, and affection as warmth. One study has explored the metaphor of thoughts themselves as physical entities. Consider *grasping* a concept or *chewing on* an idea. (Or, earlier in this paragraph, the *structure* of thought being *grounded* in experience.) The experimenters found that people understood phrases such as *push the argument* faster after making a relevant physical movement, such as pushing.

Wherever the soul sits, that chair is a pilot's seat. The soul functions to direct our thoughts and our behaviour, like a little man with a joystick. But if the soul has no physical qualities, no mass or might, it couldn't push that joystick; it couldn't influence our physical brain and move our limbs. To bridge the gap between them, mind and body need some shared properties, some common causal currency. The soul, as conceived, has muscle. And apparently it flexes its biceps from beyond the grave: some force is required to brush those wind chimes or move that Ouija pointer.

How much heft does the soul have? The physician Duncan Mac-Dougall famously performed a weigh-in on the soul's way out. After placing several dying patients on a scale and watching for abrupt changes during their final exhalations, MacDougall got one clean reading and reported in 1907 that this man's soul weighed about twenty-one grams, or a little more than a 'fun-size' Snickers.

We allow departed souls more than just location, strength, and mass. We often let them keep their old looks too. Because offline social cognition presents absent parties as their normal embodied selves, way offline (as in off-world) social cognition does too. We imagine spirits flying or ambulating, appendages intact. Mitch Hodge, the philosopher who studies afterlife beliefs, argues that while we have the capability to think dualistically, it's not our default. He notes that millennia of mythology and iconography depict the spirits of the dead as suspiciously indistinguishable from someone you might encounter on the street, save their period garb.

And if you can't fully take the body out of the soul, you can't take the soul out of the body either. If we really thought of corpses as defunct vehicles left by the roadside, we'd dispose of them unceremoniously. But instead we put them in their 'Sunday best' and bury them under decorative headstones that we then visit, or we stuff them and place them on the mantle (more likely for Mister Whiskers than for Uncle Jim), or we cremate them and launch them into space, as we did with Gene Roddenberry and Timothy Leary, who were each far out in their own way, and, finally, in the same way. (Those on a smaller budget can turn cremains into diamonds by LifeGem or concrete reefs by Memorial Reefs.) And sometimes we take organs from the dead and place them in new recipients, who then claim to take on personality characteristics of the previous owner, as we saw in chapter 1.

Some people believe a healthy corpse is required for a healthy afterlife. Ancient Egyptians preserved the bodies of nobility through mummification, even entombing them with supplies they might need in the hereafter. And on *The Daily Show*, Jon Stewart recounted his grandfather's preferred burial orientation with regard to his grandmother: 'He was very clear that she should be on the right side, because just to the left, a doctor was buried, and he wanted very much to be buried next to a doctor just in case he needed anything.'

On the topside of the turf, we often judge ourselves and others based on appearances. Who hasn't felt embarrassment regarding a birthmark or complimented a date on her big brown eyes? As if congenital features are something someone can take credit or blame for? As if flesh is the soul manifest? And consider the shame surrounding deformity, disease, and body malfunctions such as incontinence, as if these corporal failings signified moral failings.

We frequently treat our bodies not as objects but as subjects: you don't *have* a body; you *are* a body. This equation lines up especially when we feel most alive, when we're expertly interacting with the world. Take that away and you feel diminished. George Dedlow, the fictional Civil War quadruple amputee, reported, 'I found to my horror that at times I was less conscious of myself . . . I thus reached the conclusion that a man is not his brain, or any one part of it, but all of his economy, and that to lose any part must lessen this sense of his own existence.' People sometimes say, 'You are not your illness', but in a very fundamental way you are.

My own identification with my body has had its ups and downs. In school I was diagnosed with scoliosis, an asymmetric curvature of the spine. You don't notice it on casual observation, but it strikes a clear S shape on an X-ray, and I wore a painful back brace at night for a couple of years. Just knowing the deformity was there under my skin hit me hard, and the insult cut to the bone, and deeper. People use spine-related metaphors for character: *he's twisted, she's spineless, he's an upright guy*. I felt defective to my core.

My body image has since improved, and my self-esteem along with it. I was not a particularly confident athlete growing up, but as a teenage raver I found embodied expression in dance. Despite my twisted spine, I now feel agile, robust, even graceful, inside as well as out. To use Dedlow's term, my economy has expanded.

Living in Denial

In the late 1990s, a couple of American teenagers began planning a big school project, something they hoped would not only cause a local ripple but put their town on the map. For months, they documented their vision and their preparations in journals and videotapes. 'We're going to kick-start a revolution', one says in a recording. 'Directors will be fighting over this story', the other adds.

On 20 April 1999, Eric Harris and Dylan Klebold taped their final good-byes and walked into Columbine High School, near Denver. An hour later fifteen people were dead, including Harris and Klebold, who'd turned their guns on themselves.

What makes some people decide to take themselves out in a blaze of glory, the light from which they can't even bask in? What's the point of making a name for yourself if it requires erasing the person behind the name? 'Isn't it fun to get the respect we're going to deserve?' Harris asked. Even though 'we're going to die doing it.' From suicide bombers to kamikaze pilots, how do organisms programmed for survival and reproduction come to value abstractions such as respect and honour over their own genetic imperative?

||||||||||||||

In a segment produced by the satirical Onion News Network, scientists reported they had finally taught a gorilla that it will someday die. 'The researchers say at first Quigley could only communicate rudimentary fears about his own death, but he soon moved on to expressing more complex emotions like indifference and self-hatred', the reporter explains. We see captions under the gesturing gorilla. First, 'Sad. Cry. Scared.' Then, 'Stupid Quigley. Now sees truth. Existence. Cruel joke.' Eventually, the primate suffers a panic attack, crying in anguish and banging his head against the wall. The successful researchers beam with pride and move on to their next targets of enlightenment: two adorable bunnies.

The gag gets to the heart of existentialism. Unlike every other animal on the planet, humans recognize their own mortality; we're aware that our Earth-bound lives are temporary blips in the grand scheme of the universe – that everything we ever work for will eventually be taken from us, and we will rot away, returning to the void from which we emerged. So the question is: granted the ability to diagnose our certain annihilation, how the hell do we hold our shit together enough to care about things like matching placemats or recording *Celebrity Big Brother*? Why don't we lose it like Quigley?

As discussed above, most people believe in an afterlife, and even atheists can't fully imagine the end of consciousness, but that's not enough to nullify our neuroses. We clearly still fear death – strongly. And it's one of the few certainties in life. So why doesn't it haunt our every step, nagging us to crumble in the face of its overbearing promise of ablation? Because the same sophisticated cognition that lets us see the dead end also lets us think our way out of it. We all harbour a belief in what psychologists call *symbolic immortality* – we have a sense that some part of us will survive through our legacy or our joint identity with something larger than ourselves. Even as your biological self disintegrates, your symbolic self – comprising your name, your identity, your ideas – soldiers on.

Symbolic immortality prevents existential anxiety in a way that belief in an afterlife does not, because while your thoughts don't often alight on the pearly gates in everyday matters, nearly everything you do relates to your symbolic self and has the opportunity to bolster it, to fortify it for the long postmortem haul. 'Mankind's common instinct for reality . . . has always held the world to be essentially a theatre for heroism', William James wrote. Heroism need not involve saving a baby from a burning building; heroism in this context refers to anything that contributes to one's legacy, however trivial.

The cultural anthropologist Ernest Becker wrote in his Pulitzer Prize–winning book *The Denial of Death*, 'It doesn't matter whether the cultural hero-system' – the societal structure – 'is frankly magical, religious, and primitive or secular, scientific, and civilized. It is still a mythical hero system in which people serve in order to earn a feeling of primary value, of cosmic specialness, of ultimate usefulness to creation, of unshakeable meaning . . . In this sense everything that man does is religious and heroic.' It is our contributions of ourselves to the symbolic world of meaning that assures us permanence beyond burial. But for putting your name in a Hall of Fame to reduce the sting of death, as if your very consciousness would reside on those plaques and in those fans' memories: while perfectly natural and perhaps essential for sanity, this is a leap that I consider magical thinking.

Becker's work, which illuminated and elaborated on the ideas of earlier writers, including Sigmund Freud, Søren Kierkegaard, and the psychoanalysts Norman Brown and Otto Rank, didn't receive much attention from experimental psychologists at first. Existentialism perhaps seemed too hard to tackle in the lab. Then in the early 1980s three early-career researchers discovered *The Denial of Death* and refined Becker's conjectures into a set of testable hypotheses. Because Becker referred to the 'terror' that mortality arouses in us, Jeff Greenberg, Tom Pyszczynski, and Sheldon Solomon labelled the system of hypotheses 'terror management theory', and today hundreds of published studies have supported TMT's predictions. The ways in which we manage the terror of death have been shown to influence such disparate areas as romance, social prejudice, self-esteem, art appreciation, consumerism, religion, health behaviour, politics, altruism, and sex. Greenberg tells me TMT's reach is 'so broad that if taken fully seriously it could require restructuring introductory psychology texts'.

At its core, TMT proposes that managing the terror of death requires two elements: belief in a coherent and stable cultural worldview, and belief that one is valued within this worldview. Combined, they place you as part of something larger than yourself, something that will continue beyond your departure and carry a piece of you with it.

Most TMT experiments test what's called the *mortality salience hypothesis*, the idea that any existing psychological defences against the threat of death should be especially high after subtle reminders of our mortality. In a typical study, half the subjects jot down their thoughts about what will happen when they die, and a control group writes about dental pain or public speaking or some other anxiety-inducing topic that doesn't involve death. Then they perform a brief filler task to distract them so that death thoughts are relatively accessible but not the focus of awareness. (Alternately, mortality salience can be induced by subliminally flashing 'death' on a computer screen or testing subjects in the vicinity of a funeral home. The idea is to amplify the automatic defences against death anxiety that are constantly operating in the background rather than to activate more conscious defences.) Finally, subjects rate their attitudes towards a person or a point of view.

Increasing mortality salience tends to enhance one's support for views similar to one's own, a reaction called *worldview defence*. Alternative views challenge the cultural foundation on which one's identity is grounded, and so we denounce them. In an early demonstration of worldview defence, mortality salience enhanced Christians' positive judgements of other Christians and reduced their opinions of Jews. In another early test, city court judges reminded of death upheld their moral principles by recommending a higher bond for a hypothetical prostitute compared to judges in the control group – $455 (about £285) versus $50 (£30). If our cultural values – the standards by which we measure our heroism – are threatened by a heathen or a whore, we must fight back or risk returning to dust unremarked! (So the subconscious snarl goes.) According to Becker, 'people try so hard to win converts for their point of view because it is more than merely an out-look on life: it is an immortality formula'.

With our worldview confirmed – with infinity's formula defined, the rules for redemption spelled out, the hall of fame's frame erected – we can vie for our spot in that hallowed hall, our place in history. The reward for living up to an established system of values is immortality. Or so we believe. The real (and immediate) reward is the feeling of self-esteem, the impression that you're doing things right and deserve respect. So self-esteem and a sense of symbolic immortality are really two sides of the same coin, according to TMT. Terror management theorists argue that high self-esteem functions partially, if not mainly, as a buffer against existential anxiety. Studies show that mortality salience increases attempts to boost self-esteem – by, say, taking credit for success and denying blame for failure (more than we do normally). Further, inflating self-esteem reduces the effect of mortality salience on worldview defence; if you think you're the bee's knees, you needn't fret over your legacy's fragility.

According to the theory, our defences against death anxiety mobilize nearly every aspect of our symbolic selves, influencing areas of our lives that have no clear connection to death. For example, mortality salience reduces appreciation of abstract modern art in people who enjoy order and routine; random paint blobs interfere with their construction of a meaningful reality. Mortality salience increases one's commitment to a romantic relationship; we apparently include significant others within our expanded self-concepts to deny our individual vulnerability to annihilation. And people naturally high in both existential anxiety and materialism (as in consumerism, not atheism) feel stronger identification with their favourite brands than others do, which may be their way of seeking meaningful connections with fellow brand users. Give an Apple fanboy a hug today.

All these behaviours occur below the radar of conscious reaper awareness. In multiple realms of life, the denial of death can't be denied. And this denial, this identification of ourselves with the ripples we leave in the world, is magical thinking.

Acts of Heroism

In the 1970s the American psychiatrist Robert Jay Lifton outlined five modes through which we obtain symbolic immortality, the most important being *biosocial* and *creative*.

We see the biosocial mode most clearly in parenthood: the desire to live on through our children. In studies, mortality salience increases the desire to have children and the accessibility of offspring-related thoughts. And while thoughts of death typically lead to negative stereotyping of other cultures, thoughts of procreation reduce such mortality-inspired worldview defence. You may not think of your two-year-old as anxiety reducing, but she does prevent the existential kind.

The biosocial mode of symbolic immortality does not restrict itself to genes. Evolution dictates the wide distribution of our DNA, but on a psychological level we may care more about the spread of ideas. Parents often model their children after themselves, raising them to like the same music and espouse the same politics and fear the same gods, sometimes with more success than others.

The biosocial mode doesn't restrict itself to children either. One can live on through any kind of social affiliation, whether the spiritual prosthetic is your family, a lover, your nation, a favourite sports team, or humanity in general – anything with a chance at survival beyond oneself. In a questionnaire administered before the first football game of the season at the University of Arizona, mortality salience increased students' identification with and optimism about the team. In the face of death, they wanted to be part of a larger enterprise and wanted to believe that this vehicle for permanence would serve them well. A week later, after the team's loss, mortality salience shifted students' affiliations from the team to the basketball team. No one wants to hitch his soul to a sinking ship.

Famous people also serve as buoys, and we do our best to keep them afloat. In an unpublished study, mortality salience increased the span of time Americans thought a number of cultural icons would be remembered after their deaths. The more representative of American values, the longer the symbolic life extension. While Tom Cruise's reign of memorableness jumped from 54 to 71 years in the face of death, Oprah's went from 77 to 140, and President Abraham Lincoln's from 755 to 4792. And we convince ourselves that symbols of our identity approach immortality not just figuratively but literally. Turkish subjects deemed a plane less likely to crash if it carried a celebrity. The more the passenger represented Turkish values, the greater his protective force on the aircraft. According to Otto Rank, 'Every group, however small or great, has, as such, an "individual" impulse for eternalization, which manifests itself in the creation of and care for national, religious, and artistic heroes.'

The second major mode of symbolic immortality, the creative mode, encompasses all the ways we leave a mark on the world through our unique actions. They could be artistic, scientific, athletic, sartorial. According to existential psychologists, I'm writing this very book in part to memorialize myself in libraries and digital archives. No one will ever be able to deny I was here, or, more to the point, deny part of me is still there, in the future.

Feelings of self-worth and self-permanence require much less than the erection of a great pyramid or an eponymous grand theory. According to Lifton, a therapist's or physician's treatment of a patient creates ripples tickling posterity. He goes further: 'These issues are germane to more humble everyday offerings of nurturing or even kindness – in relationships of love, friendship, and at times even anonymous encounter. Indeed, any form of acting upon others contains important perceptions of timeless consequences.' A simple smile at a passing stranger receives notice, rendering it, in its own small way, heroic, and immortalizing.

But do bids for immortality truly motivate any of these creative acts? Maybe I'm writing this book purely for the chance to announce myself as an 'author' at fancy dinner parties and the occasional drive-thru window. Maybe I don't give a whit whether all copies go missing the day I die. Actions aimed at an audience that has survived us can be tough to tease apart from actions meant to benefit our premortem reputations.

Clearer cases of attempts at symbolic immortality emerge when examining individuals on their deathbeds. Spending can be a way of leaving your mark, and many names end up on buildings and professorships thanks to bequests. A survey of supporters of three large charities revealed that those who'd pledged bequests had a stronger 'need to live on': they were more likely to confirm, for example, that 'I would want to be remembered for having supported [this organization].' An exemplary quote from a donor in a focus group: 'Someone, somewhere will know this was something that mattered to me.'

As Paul Edwards, the philosopher (and editor of Bertrand Russell's volume *Why I Am Not a Christian*), has written, 'Even existential philosophers . . . appoint literary executors.' One might predict that with no afterlife awaiting them, *especially* existential philosophers would appoint literary executors. They want *something* of themselves to carry on.

Dying to Live

We've established that fear of death leads to worldview defence and self-esteem bolstering, purportedly in the name of symbolic immortality, but does what we're calling a sense of symbolic immortality succeed in fending off fear of literal death? Is symbolic immortality similar enough to literal immortality to deserve the title *immortality*?

After all, Woody Allen once remarked, 'I don't want to attain immortality through my work; I want to attain immortality by not dying.'

The most direct evidence comes from a study in which people rated their agreement to phrases such as *After my death, my life will still have meaning through the lives of my children* and *After my death, my life will still have meaning through the things I have made or created*, providing a measure of each subject's sense of biosocial and creative symbolic immortality. Those who scored highly showed reduced fear of several elements of death, including decomposition of the body, uncertainty of what to expect, and loss of self-fulfilment. Knowing they would carry on through symbols – memories and creative works that stand in for them – meant they didn't worry so much about not carrying on in flesh and blood.

Despite this work, one can still wonder whether a sense of symbolic immortality actually reduces the belief that one's existence will end – whether it's a form of magical thinking – or whether it just helps one cope with that belief. One might then ask whether it's even possible to differentiate experimentally the former from the latter interpretation. I put the issue to Greenberg. His response:

> I don't know an empirical way to answer it, yet. I lean towards the former in this sense. To the extent that one can remain embedded in a largely illusory symbolic reality in which we are people with roles and achievements and collective identifications, then we can believe we will live on past our physical death. If I am Jeff Greenberg, co-originator of TMT, I will still be that after I have died. So it can provide a powerful sense of continuance, just as Lincoln and Marilyn Monroe seem as part of reality now as people who are currently living (if not more so).

Indeed, Elvis earned $60 million – £40 million – in 2009. He doesn't need to be in the building to make bank.

So if people treat postmortem notoriety as the functional equivalent of an afterlife, which is really just a modified extension of regular life, we can see how the Columbine shooters would trade time on Earth for a biopic directed by Quentin Tarantino. (Harris and Klebold actually debated whether Tarantino or Spielberg should handle their cinematic eulogy. I hope they're Michael Moore fans.) Their double suicide wasn't a surrender; it was a swap: one kind of life for another.

Many of us would make a similar swap, although under different banners and with fewer casualties. People have given their lives for individual fame, or in support of larger groups they identify with. In one study, reminding British students of their mortality increased their motivation to sacrifice their individual selves for the sake of a collective self: death salience made them more likely to say they would die for England or that the continuation of the British way of life took precedence over their personal safety.

The researchers who conducted the study, Clay Routledge and Jamie Arndt, point out less cut-and-dried trade-offs between life and symbolic immortality: volunteering for your country's military, fasting in protest of a government policy, or serving as a missionary in an unstable region. You're risking your body to maintain your soul's getaway ship, whether it's a nation, an ethnic group, or a church. Other studies they cite indicate that self-esteem also comes before safety when faced with mortality: reminders of death increase risky driving in people who especially value driving and tanning in those who especially value beauty.

Still, if your goal is to nurture a particular establishment and then endear yourself to that establishment for the sake of personal enshrinement – to achieve symbolic immortality through worldview defence and status striving – then bringing pipe bombs to school seems misguided. You'll not earn many fans. But heroism requires individuality. People like rebels, especially in America, and the Columbine boys had a taste of the revolutionary in them. Harris wrote in a journal that those around him were 'robot[s]'. The ideal balance between individuality and conformity is called *optimal distinctiveness*; standing out too much makes you an outlier ripe for dismissal, but fitting a mould too precisely reduces you to a forgettable automaton. Harris and Klebold knew they were making a dramatic statement, but they didn't think they were going so far off the beaten path that no one would pay them any heed. According to Klebold, 'I know we're gonna have followers.'

Unfortunately, they did. Before killing thirty-two people and himself at Virginia Tech in 2007, Seung-Hui Cho paid his respects in a videotape to 'martyrs like Eric and Dylan.' Surely, they would have considered such an homage a testament to their immortality. We can only hope that over time their ghosts serve not as continuing agents of torment but as reminders of our duties to help those in pain.

Gods with Anuses

'No mistake – the turd is mankind's real threat.' Becker could be referring here in *The Denial of Death* to the spread of disease in developing areas due to inadequate sewage disposal, or describing a fictitious supervillain named The Turd who stomps cities flat and leaves nations besmirched. But no, he's claiming that the lonely turd sitting in a toilet bowl can arouse mortal terror – terror worse than any giant monster, because it's immune to heroics.

To achieve symbolic immortality, we must live a life of meaning; we must fashion a symbolic existence. So we define ourselves by our thoughts and aspirations and memories and intelligence and morality, things that can be transmitted through culture as abstractions. We extend our identities beyond the natural world; we become supernatural. But there's another side to our existence, one less angelic. It's the side that shits and bleeds and oozes and breaks and eventually stops working and rots away, just like every other creature on the planet. In Becker's terms, we are 'gods with anuses'. And this fact tears us apart; to the degree that we are mortal animals, we are not immortal souls.

Becker really gets going when discussing the conflict between our embodiment and our ethereality, and how turds are the poignant, pungent essence of that problem:

> Excreting is the curse that threatens madness because it shows man his abject finitude, his physicalness, the likely unreality of his hopes and dreams. But even more immediately, it represents man's utter bafflement at the sheer *non-sense* of creation: to fashion the sublime miracle of the human face, the *mysterium tremendum* of radiant feminine beauty, the veritable goddesses that beautiful women are; to bring all this out of nothing, out of the void, and make it shine in noonday; to take such a miracle and put miracles again within it, deep in the mystery of eyes that peer out – the eye that gave even the dry Darwin a chill: to do all this, and to combine it with an anus that shits! It is too much. Nature mocks us, and poets live in torture.

We would love to see ourselves as immortal beings, but faeces is the sludge in the mechanics of magical thinking.

Several studies have supported the hypothesis that reminders of our similarity to other animals make thoughts of death more accessible. Jamie Goldenberg, a psychologist at the University of South Florida, has led the charge in this area. In one experiment, subjects read an essay stating, in part, 'although we like to think that we are special and unique, our bodies work in pretty much the same way as the bodies of all other animals', or one stating, 'we are not simple selfish creatures driven by hunger or lust, but complex individuals with a will of our own'. Then they were asked several questions about speaking in public, or several questions about how bothered they are by unflushed toilets, vomit, and people clearing their throats. Goldenberg found that afterward, subjects who had been told that humans are like animals and then made to think about disgusting bodily effluvia had death on the mind more than the other subjects did. Their own embodiment became an existential threat.

Not just any reminder of our physicality makes us think of death, however. For example, reading the essay emphasizing our similarity to animals without the additional graphic pictures doesn't increase existential angst in most people. Perhaps that's because we find ways to make most elements of our time between dust and dust life-affirming and postlife-extending. We wrap our actions and bodies in meaning. We express ourselves through our clothes and our physiques and our manners. People conquer and symbolize their bodies through yoga and make-up and shaving and chewing with their mouths closed. Some cultures have used foot binding and lip plates and neck rings. As Simone de Beauvoir wrote, 'A woman is rendered more desirable to the extent that nature is more highly developed in her and more rigorously confined.' Through these acts, we separate ourselves from the slovenly, hairy creatures who don't tame and tailor their embodiment to express refinement or creativity or irony. We demonstrate that we are more than spiritless slaves to nature, that we can rise above it into the supernatural realm of symbols. That we are defined by the magic inside us, not the flesh we temporarily inhabit. Through a complex combination of treating our bodies as instruments of expression and denying our identity with them, we highlight the transcendent aspects of our humanity and intensify our sense of symbolic immortality.

Some of society's most rigid prescriptions, bodily or otherwise, involve making sex sanitary or symbolic, subjecting this animalistic act to taboos or transforming it into a language of personal intimacy – taming the beast with two backs. Goldenberg has demonstrated the psychological threat of copulation. In one study, reading an essay comparing humans to animals and then thinking about the physical aspects of sex ('the smell of sex', 'having an orgasm', 'a tongue in my ear') made people more likely to think about death. Reading the essay and then reviewing the romantic elements of sex ('feeling close', 'connecting spiritually') did not. Love puts the beast on a leash. In another study, reading the humans-are-animals essay and then thinking about death reduced the appeal of sex's physical side. Mortal disgust at the gooey procedure of procreation may significantly contribute to the popularity of candles and corsets, which enshrine our sweaty slavery to our loins as a sacred savouring of the erotic arts.

And sometimes our deepest aversions aren't simply ignored in the heat of the moment or in the service of a higher romantic purpose; sometimes they become transformed into specific attractions. Becker iterates an argument that fetishes serve to reduce anxiety about human creatureliness: 'The fetish takes "species meat" and weaves a magical spell around it.' Is it any wonder that one of the most fetishized body parts – the foot – is also one of the most disgusting? (Becker calls feet 'the absolute and unmitigated testimonial to our degraded animality'.) It's not quite clear how the alchemy from mortal flesh to magical charm operates, but in any event it provides anxious individuals the courage to perform the sexual act.

There are a few possible explanations for why men fetishize body parts more than women do. One is that males typically initiate sex, so they need to sublimate their fears to a greater extent. Another is that women's more intimate involvement in reproduction, from menstruation to pregnancy to lactation, makes them particularly strong reminders of humanity's biological basis, so their bodies require greater cultural confinement. (Indeed, breasts and genitalia are both covered by women and coveted by men. And in some parts of Africa, women inflict 'genital curses' on men by exposing themselves; their reproductive organs and fluids have magical polluting power.) Finally, in a male-dominated society, boys may have learned that they must tame all aspects of nature, including the womanly other. (Although many women now prefer that men, too, trim their pubic hair.)

While humans can dress up fornication (ahem, making love) and ingestion (dining), some biological necessities such as defecation resist cultural decoration. You can't put a pretty bow on a turd. Well, technically you can, but you'll receive some odd looks. So we resort to shame and denial. This is why toilet stalls have doors and languages have euphemisms.

Religious leaders have made special efforts to separate their saints (and themselves) from the trappings of creatureliness. Valentinus, the second-century Gnostic, wrote that Jesus 'ate and drank in his own manner, without excreting food'. The prominent Puritan minister Cotton Mather wrote in his journal three centuries ago,

> I was once emptying the *Cistern of Nature*, and making *Water* at the Wall. At the same time, there came a *Dog*, who did so too, before me. Thought I; 'What mean and vile Things are the Children of Men, in this mortal State! How much do our *natural Necessities* abase us and place us in some regard, on the Level with the very *Dogs!*'

Even atheists have their sacred cows. The Russian-born psychologist Eugene Subbotsky wrote in his book *Magic and the Mind:* 'The God-like figure of Stalin in school textbooks is still before my eyes. When I was 6 (it was 1954), I remember discussing with other children the issue of whether Stalin had to go to the toilet, and most of us were of the opinion that he did not.' And I recall a conversation from school in which guys were discussing whether the hottest girl in school pooped. 'Okay', one relented, 'I guess she does. But they're probably perfect little pellets.'

This brings us back to fetishes. (Or rather *I* bring us back to fetishes. Caught red-handed.) If poop is so gross, and fetishes are defences, why aren't there more coprophiliacs (people turned on by poop)? Probably in part because faeces usually stays out of sight, so there's no reason to overcome the disgust. But during bedroom activities we frequently come eye-to-brown-eye with the orifice responsible for it, and bingo, there's another fetish: anal sex. By treating a sewage spout as a porthole to pleasure, the rectum becomes a tunnel to magical thinking.

(I don't mean to imply that behinds and bare feet offend us solely because they trigger abstract concerns about our animal mortality. They also harbour dangerous and malodorous bacteria; our disgust signals a very concrete biological threat.)

Weaving the web of life into a cultural fabric creates a Catch-22.

By treating the drab, sometimes disgusting, occasionally savage, and generally absurd facts of our daily existence as meaningful contributions to our heroic narrative, we can live with them. We can function knowing that ambulating around the planet, incorporating and depositing pieces of surrounding matter, and occasionally producing new bundles of meat that resemble us is somehow more than just a story about atoms. It's a story about souls.

But at the same time wrapping ourselves in a cultural fabric confines us, insulates us from the act of actually living. Goldenberg has found that in our defences against awareness of human animality, we enforce unhealthy ideals of beauty, we avoid proper medical attention, and sometimes we even deny ourselves pleasurable physical sensations. Ascetics have no fun at all.

To live fully, we must embrace our complete humanity – the spiritual and the material. To be human is not merely to exist on an abstract plane of symbols; it is also to be in the trenches, to have orgasms and bowel movements, hopefully not at the same time, and to experience hunger and thirst and pain as well as ice cream and rain on your face and tickle attacks. The physical plane may not be perfect, but we're all in this together, and to deny unsanitized reality would be to live a life of isolation and paralysis. Try a tongue in your ear just once.

Heaven on Earth

In *A Christmas Carol*, Ebenezer Scrooge receives the ghost of his long-dead business partner, Jacob Marley. Marley haunts the world in chains, forever burdened with the miserly misery he created for himself in life. He threatens the same fate for Scrooge and summons three additional spirits to aid his case.

Scrooge changes his ways overnight, joining his nephew for Christmas dinner the next day and sending a prize turkey to the home of his underpaid clerk, Bob Cratchit. But one could argue that it wasn't the prospect of an unpleasant afterlife that opened his heart and his purse. It was his apparition-induced recognition of the wants and joys of his living companions – including Tiny Tim, whose precarious young life illuminated those around him – and an emotional investment in this life for its own sake. To seal the deal, the Ghost of Christmas Yet to Come concluded the night's medley with a visit to a manifestly unattended 'Ebenezer'-engraved tombstone. Staring at your own grave has a way of clarifying priorities.

While our defences against existential angst can escalate the objectification of women, prejudice against people with opposing beliefs, and behaviours that risk our own health, they can also lead us to live our lives according to constructive social values. In one study, subjects expressed more favourable attitudes towards charities when stopped in front of a funeral home than when stopped three blocks away. The researchers called it *the Scrooge effect*. Describing our attempts at symbolic immortality, the sociologist Michael Kearl has written, 'like the art of karate where one uses the energy of one's adversary, the power of death is diverted to personal growth and social development'. We invest in our own supernatural significance in the face of nature's necessities.

But we're left with another paradox. In some respects death makes life meaningless in its finitude. Shakespeare famously wrote,

> *Life's but a walking shadow, a poor player,*
> *That struts and frets his hour upon the stage,*
> *And then is heard no more. It is a tale*
> *Told by an idiot, full of sound and fury,*
> *Signifying nothing.*

And yet, because we value scarce resources, death is what makes life so precious. In the film *Fight Club*, Tyler Durden (Brad Pitt) holds a gun to the head of a convenience shop clerk, then later says to his companion, 'Tomorrow will be the most beautiful day in Raymond K. Hessel's life. His breakfast will taste better than any meal you and I have ever had.'

The psychiatrist Irvin Yalom wrote in his book *Existential Psychotherapy*, 'Although the *physicality* of death destroys man, the *idea* of death saves him.' In a laboratory study, after subjects found death-related words in a word-search puzzle, they more easily found meaning in life. And in clinical accounts, people feel more authentic and more appreciative of small pleasures after brushes with death. The palliative-care doctor David Kuhl has written, 'For many, the diagnosis of a terminal illness or the experience of the illness serves as a roar of awakening.'

Many of us will never experience this roar, but we can infuse our lives with a constant growl. The idea is to face mortality as frankly as you can without freaking out. To accept a manageable share of anxiety and to channel it towards building a heaven here on Earth. Unfortunately, there's no how-to manual for this – neuroses and nirvanas are as individual as fingerprints. Maybe you find meaning in your family, your work, or your hobbies. You're on your own to improvise. Just find courage in the thought that real heroes don't need scripts.

And they don't need sequels either.

6

||||||||||||||||||||||||||

The World Is Alive

Animals, Objects, and Gods Are People, Too

I'm kneeling next to Wes Richardson, a resident at the US Veterans Affairs Medical Center in Washington, DC. Seated in a wheelchair and wearing a green polo shirt, striped pyjama bottoms, and a 101st Airborne baseball cap, he's telling me a story, and my attention is rapt.

'I'm used to rabbits. I used to raise rabbits. White rabbits. Easter rabbits. One of them came, a jackrabbit. Had *huge* ears. They would stick straight up in the air. He would fight you, too. He was dangerous. Couldn't let the little kids mess with him. That's the jackrabbit. But the other rabbits, you know. But then they would run away. Then I had a dog. He was a fighter, like a pit bull. He wasn't a pit bull. He was a German shepherd and a boxer. He was a killer dog. I'm serious. Named him Rocky. . . .'

Okay, so it's not *Lassie*. But the story is notable if only for the fact that Richardson is recovering from a stroke and only a few months ago didn't speak at all. Now he's painting pictures with words for a stranger.

There's something else worth mentioning. As Richardson speaks, he's petting a $6000 (£3750) robotic baby seal. And the seal may have something to do with his ability to tell a story.

The robot is manufactured under the name Paro (short for personal robot) and sold to hospitals and nursing homes around the world. Takanori Shibata, its Japanese inventor, recognized the benefits of animal-assisted therapy – pets lower blood pressure and reduce depression in their owners – but many facilities don't allow animals, so he set out to find a solution. The result is a three-kilogram robot modelled after a baby harp seal.

Paro can move its neck up and down and side to side, it can move its flippers and its tail, and it can blink its eyelids. It's sensitive to light, sounds, and touch. And it plays recordings of real baby harp seals. When you hold Paro in your arms and stroke its soft white fur, it squirms around, looks at you, and makes appreciative bleats. Your heart melts.

Wes Richardson might have started talking without Paro's help, according to Dorothy Marette, a clinical psychologist at the medical centre and the caretaker of this Paro (or 'Fluffy', as many have come to call it), but when his words came back, Marette says, 'he was speaking to Paro rather than to people'.

Richardson has become more active in other ways. 'Before, he was in his room all the time,' Marette says. 'He wasn't out at all. Now he's in the dining room every day eating, he's out watching TV, he participates in some of the exercise groups.' Again, maybe he would have come out of his shell anyway, but the seal seems at least a likely enabler.

Richardson is not out of touch with reality. He realizes he's holding a robot. 'It's amazing how somebody can make this,' he tells me. But his first reaction to seeing Paro on my visit was a greeting, to the robot: 'What's up?'

||||||||||||||

People will treat anything as a person. We name our cars, shout at tables when we bump into them, give birthday gifts to our pets, call plants 'thirsty', cajole clouds to part, and pray to gods for forgiveness. We act as though every encounter were a social one, as if the world were populated with minds like our own, as though the universe itself were conscious. 'We find human faces in the moon, armies in the clouds; and by a natural propensity, if not corrected by experience and reflection, ascribe malice or good-will to every thing, that hurts or pleases us', David Hume wrote some 250 years ago. These seem the symptoms of a madman, and yet they describe the mind of a healthy human. Wherever we go, we are not alone.

Treating something not alive as if it's alive is called *animism*. Treating something nonhuman as if it's human is called *anthropomorphism*. The two often go hand in hand, as animism feeds into anthropomorphism: when we treat something as alive, we treat it as *alive like us*, with our full catalogue of mental states.

The rich variety of our animism and anthropomorphism really comes to light in art, idiom, and metaphor, where we not only don't suppress our habits but amplify them for effect. 'Hence the frequency and beauty of the *prosopopoeia* in poetry; where trees, mountains and streams are personified, and the inanimate parts of nature acquire sentiment and passion', Hume wrote.

The anthropologist Stewart Guthrie provides a comprehensive review of our propensity to personify in his 1993 book *Faces in the Clouds*. In documenting examples from popular culture and everyday life, he notes, for instance, that gardeners speak of what conditions their plants *like*, mechanics talk about *cranky* engines, sailors refer to ships as *she*, and rescue workers battle *raging* fires. Abstractions have human qualities, too: time *waits* for no man, especially with death *at your doorstep*.

Even scientists, who aim to look at nature from a distance, project themselves onto their work. Astronomers refer to the *birth* and *death* of stars, neuroscientists describe the way cells *talk* to each other, physicists remind us that nature *abhors* a vacuum, and Darwin described natural *selection* ('the selecting power of nature, infinitely wise').

Piaget noted animism in children, but as we age we don't grow out of our childish magical thinking. We learn to correct for it. At heart, we still live in an enchanted kingdom where spirits inhabit everything around us, where the wind literally howls, the brooks literally babble, and a guardian in the sky looks after us.

Anthropomorphism is worth understanding 'not just because it tells us about why people dress up their pets or talk to their plants, but it's really about how people understand minds more broadly', Adam Waytz, who has studied the phenomenon extensively, told me.

And it has implications for animal rights, parenting, product design, entertainment, marketing, human-computer interaction, environmentalism, and religion. It's a promiscuous and powerful instinct.

So hey, anthropomorphism: what's up?

Seeking Minds

At the top of my field of vision is a blurry edge of nose, in front of me are waving hands. . . . Around me bags of skin are draped over chairs, and stuffed into pieces of cloth, they shift and protrude in unexpected ways. . . . Two dark spots near the top of them swivel restlessly back and forth. A hole beneath the spots fills with food and from it comes a stream of noises. Imagine that the noisy skin-bags suddenly moved toward you, and their noises grew loud, and you had no idea why, no way of explaining them or predicting what they would do next.

This unnerving scene is not from a horror film or Franz Kafka novel. Concocted by the psychologist Alison Gopnik, it's a typical family dinner, as someone who does not see minds even in other people would experience it. Without the ability to attribute beliefs and desires to those around us, the world would be a confusing place indeed. Just a jumble of random movements by skin bags stuffed into cloth.

We see minds in things (magical thinking) because we've evolved to see minds in people. Recognizing minds opens up another dimension of reality and sets humans apart from most of the rest of the animal kingdom. This capacity for mentalizing, sometimes called *theory of mind*, provides enormous benefits to those who wield it. Inferring mental states in other people allows us to understand and predict their behaviour. And awareness of mental states in ourselves allows for self-reflection and executive control. Once we enter the world of social strategizing, mindsight becomes our natural mode of navigation. As the cognitive scientist Dan Sperber has put it, 'Attribution of mental states is to humans as echolocation is to the bat.'

In the last few years, three psychologists from the University of Chicago – Nicholas Epley, Adam Waytz (now at Harvard), and John Cacioppo (I'll call them EWC) – have set an agenda for the study of anthropomorphism. They hope to explicate how mental attribution works and why we apply it to inanimate objects (as well as why we sometimes don't apply it to our fellow humans). To guide the field, EWC have organized the literature into three main factors influencing anthropomorphism. The first factor is how well we understand agents and how easily our knowledge of agents is activated. (In psych-speak, an agent is not a spy but any entity that can initiate an action according to goals – typically a person.) The second is how much we desire control over a situation; identifying agents and predicting their behaviour allows us to better anticipate the events around us, as well as interact with those agents. And the third is how lonely we are – how much we desire a companion. EWC, rearranging the factors to form a snappy acronym, call their theory SEEK, for Sociality, Effectance (i.e., control), and Elicited agent Knowledge. It earned them a Theoretical Innovation Prize from the Society of Personality and Social Psychology, and it will serve as a handy framework for organizing the bulk of this chapter.

What We Know about People

We wouldn't anthropomorphize if we didn't know so much about people. From birth, we begin collecting intelligence on agents, crawling towards an awareness of their existence, and of our own identity as one. By three months, infants can follow another's gaze, looking in the same direction as someone else. By twelve months, they're using communicative gestures such as pointing. By eighteen months they can recognize people's preferences, even if those likes and dislikes differ from their own. By eighteen months they also acknowledge goals. At two years, children are openly discussing feelings in themselves and others. And at four or five, children can explicitly reason about false beliefs, mental representations that don't accord with reality. (Some evidence suggests children are sensitive to another's point of view in false-belief tasks as young as seven months.)

All of this goes to illustrate a speedy developmental trajectory in which before we can even use the big-person potty we're making inferences about mercurial abstractions we cannot see – thoughts. By imputing minds, we're turning bags of skin into agents. Agents who can be angered, or soothed, or implored for assistance, or manipulated into buying us sweets. Other people provide comfort and protection and funny faces; they are the most important parts of our environment. Through years of continuous interaction and careful observation, we build a massive storehouse of knowledge about them.

There are debates about how we accumulate and apply this knowledge. Some researchers have argued that theory of mind resembles scientific theory in that we use an abstract body of laws to understand people's thinking. Others believe mind reading is more organic, and that the term '*theory* of mind' is misleading. They propose a strong simulation component, in which we relate to other people by implicitly putting ourselves in their shoes.

Support for the simulation account has come from the study of *mirror neurons*, discovered in the 1990s. A group of Italian neuroscientists noticed that certain cells in monkeys' brains would fire whether the monkeys performed an action or the monkeys watched an experimenter perform the same action. In observing another agent, a monkey's brain acts almost as if it's the other agent. Mirror neurons were soon located in humans, and were found to represent not just actions but emotions, physical sensations, and intentions. They help us understand what other people are doing, feeling, sensing, and planning, and they may play an important role in empathy.

Further evidence that simulation guides our mind reading (and thus anthropomorphism) comes from our relentless egocentrism. One consequence of our egocentrism is the *false consensus effect:* we tend to overestimate the popularity of our own opinions and judgements. When trying to get inside other people's heads, our own epistemic baggage encumbers our entry. We can't shed what we already know. As we mature, we become better at taking others' perspectives, but Epley has shown that adults are just as egocentric in their initial reading of others' minds as children are; rather than outgrowing egocentrism we merely learn to correct for it. Regarding the simulation versus theory accounts of mind reading, it seems we use a mixture of both; we start with a self-centred simulation of another's experience and then refine it based on what we know of the person and the situation.

So from an early age we become experts on minds – especially our own, which we use to understand others'. This rich and ready psychological expertise lays the groundwork for the anthropomorphizing of nonhuman agents. It just needs to be triggered. Anything showing signs of being an agent we treat as a person. If it thinks, it must think like us. Specifically, it must think like *you*.

Let's look at how we humanize three types of agents: biological, technological, and theological.

Animals

Pets are the prototypical example of nonhuman agents treated as people. Go on Amazon and you'll find books such as *Dogwear, Canine Couture*, and *Wearable Arf.* People constantly talk to their pets and occasionally allow man's best friend to substitute for a real man. It's hard to suppress interpreting dogs' behaviour in human terms, reading emotion into every glance and wag. Last week as my housemate and

I were making fun of her beagle, Molly, for having such a small brain, Molly incidentally bowed her head. My housemate immediately voiced Molly's apparent feelings: 'Come on, guys, stop making fun of me!'

I mentioned in chapter 4 that strict adherence to the definition of magic offered in the introduction would mean that attributing consciousness to the human brain is magical thinking, because consciousness is a mental phenomenon and the brain is material. Let's refine the definition to say magical thinking involves *over*attributing mental states to physical objects. So inferring uniquely human mental states in an animal brain fits the bill. Similarly, since degree of personhood sits on a spectrum (at what time during gestation do cells suddenly become a person?), I think it makes sense even to talk about anthropomorphizing humans, as we sometimes overestimate the richness of their conscious experience or the intention behind their behaviour.

To know how magical our thinking about animals is, we must first know how much mind animals actually have. Cognition in chimps and other primates is an area of great controversy, so I'll stick to the slightly more clear-cut case of dogs.

Sam Gosling, a psychologist at the University of Texas at Austin, has extensively studied 'personality' in dogs. (He uses that word despite heckling from some colleagues.) Psychologists typically measure human personality in five dimensions, referred to as the Big Five. Gosling measured analogues of four of these traits in dogs: *energy* (similar to human *extroversion*), *affection* (similar to *agreeableness*), *emotional reactivity* (for *neuroticism*), and *intelligence* (for *openness to experience*). There was no suitable approximation of the fifth dimension, *conscientiousness*. Gosling found that human judges consistently agreed on the strengths of these traits in individual dogs, and that these ratings accurately predicted behaviour in the animals. Dogs appeared to have personalities.

In my opinion, *personality* is too strong a word for simple behavioural tendencies, which is all he's demonstrated. For comparison, read how the inventor of the Paro robot describes how he's made his baby seal appear to have moods and a personality: 'Paro has internal states that can be described with words that indicate emotions. Each state has a numerical level, which changes according to the stimulation. Moreover, each state decays with time. Interaction changes its internal states, and creates Paro's character.' One can imagine mechanisms inside the dog that are just biological versions of this handful of changing numerical states. If that's all it takes to arouse our anthropomorphism of a robot, we should be careful about how much complexity we impute to the character of a canine.

Dogs may not have complicated personalities, but I will tentatively allow them rudimentary subjectivity – experiential feelings of pain, pleasure, fear, happiness, and anger. These fall into the category of *primary emotions* – states with clear behavioural signatures, short durations, early appearance in human childhood, and apparent cross-species commonality. *Secondary emotions* are more extended and reflective. These include optimism, nostalgia, resentment, and admiration. Scientists have made little serious effort to study secondary emotions in pets – mute subjects make their identification difficult – but Alexandra Horowitz, a psychologist at Barnard College, in New York City, has studied guilt in dogs, or at least behavioural displays of guilt.

In her study, owners ordered their dogs not to eat a particular treat, then left the room. Half the dogs ate the treat, and half didn't. Within each group, half the owners were told upon returning that the dog ate the treat, and half were told the dog didn't. Dogs' reactions to their owners' return were analysed. Behaviours people associate with feelings of guilt, such as lowering of the head or tail and avoidance of eye contact, were highly correlated with what the owner thought the dog had done. They were not at all correlated with what the dog had actually done. Horowitz concludes, 'What the guilty look may be is a look of fearful anticipation of punishment by the owner.' Dogs don't feel guilty; they just hear your rising voice and brace for a firm scolding, whether or not they did anything wrong.

But people don't know this. One study found that 74 percent of pet owners report that dogs can feel guilt. As for other secondary emotions, 81 percent said dogs can feel jealousy, 64 percent empathy, 58 percent pride, 51 percent shame, 49 percent grief, 34 percent disgust (really? dogs drink out of the toilet), and 30 percent embarrassment (really? see *disgust*).

While dogs may have consciousness, it's doubtful they have *self-consciousness*, an awareness of their own existence. Again, we can't ask them, but Gordon Gallup Jr has developed a test of self-awareness called the mirror test. You put a spot on an animal and see if the animal notices in the mirror that the spot is on its own body. Supposedly, if an animal can recognize itself visually, it has some concept of itself. Humans older than eighteen to twenty-four months, apes, dolphins, magpies, and elephants pass the test, but dogs don't.

If dogs don't have self-awareness, they can't have *episodic memory* either – autobiographical reconstructions of previously experienced events. They can't place themselves in another time, in another context. So they have no idea what they were doing a minute ago or what they might be doing a minute in the future. They don't ruminate or premeditate. They live reflexively, on instinct and conditioning, forever in the moment.

When we anthropomorphize things we assume their actions are intentional and have the same meaning to the actor as they do to us, which is why *Reader's Digest* gives out a Hero Pet of the Year award; they think heroism is an appropriate descriptor of a dog's instinctive or trained actions. The flip side of the potential for praise is the potential for blame. People see a lot of intelligence in dogs and thus have high expectations of their moral behaviour. We become enraged when dogs let us down, by, say, eating the cake on the counter. Didn't Fido realize from the candles that he was ruining a five-year-old's birthday party? No, animals don't know any better. In the wake of the tiger bite on the neck of Roy Horn of Siegfried & Roy, the comedian Chris Rock reminded us that 'That tiger ain't go crazy; that tiger went tiger! You know when he was really crazy? When he was riding around on a unicycle with a Hitler helmet on!' Horowitz notes that overestimating a pet's understanding of right and wrong can be frustrating and harmful for both the pet and the owner.

The question of animal intelligence is not identical to the question of animal rights, however. The philosopher of law Jeremy Bentham wrote a couple hundred years ago, 'The question is not, Can they *reason*? nor, Can they *talk*? but, Can they *suffer*?' People sometimes assume that because animals have less intellectual capacity than humans, they also have a reduced capacity to feel pain, and thus deserve fewer rights. I had an experience a couple years ago that made me at least question this line of reasoning. I was watching the roller-derby documentary *Hell on Wheels* in the cinema with my family. After a graphic leg-breaking scene, I passed out in my seat for a few seconds. Coming to, I felt severe physical discomfort – nausea, sweats – and had minimal awareness of who I was, where I was, why I felt so shitty, or even that the shittiness would ever end. This must be how animals experience suffering. When you take them to the veterinarian for a jab, they don't know it's for their own good or that it will only hurt a moment. They just feel a world of pain. It's a sad thought. I'm not going to stop eating meat, though.

The best way to get people to protect something is to make them think of it as human or at least capable of humanlike suffering. Slowing the destruction of the animals and plants in our natural environment might depend upon such a tactic. The phrase *Mother Earth* makes for good ammo in this battle. As the comic writer Jack Handey has wondered after much deep thought, 'If trees could scream, would we be so cavalier about cutting them down?' (He went on to allow that, 'We might, if they screamed all the time, for no good reason.')

Technology

Paro's white coat and meek purrs make it easy to love, but robots can get away with much less fur power in the battle for our hearts. Consider AIBO, the small plastic robotic dog manufactured by Sony that looks only nominally like a real dog. In one study, after twenty sessions with AIBO, nursing home patients were less lonely. And researchers at the University of Washington dove into chat rooms to sample owners' spontaneous descriptions of their toy pets. They found that 60 percent of forum members referred to their AIBO's mental states and 59 percent to its capacity for social rapport. One owner wrote, 'The other day I proved to myself that I do indeed treat him as if he were alive, because I was getting changed to go out, and [my AIBO] was in the room, but before I got changed I stuck him in a corner so he didn't see me!'

Okay, let's strip away the face completely and see what happens. It turns out that Roomba, the robotic hoover manufactured by iRobot, elicits emotions too. In a survey of Roomba owners, a quarter of them gave theirs a name, one in eight talked to it, one in eight dressed it up, and one in eight ascribed a personality to it, calling their cleaner silly, temperamental, flirty, or stubborn. In another study of thirty Roomba owners, eighteen felt their Roomba had intentions, feelings, and personality traits. Several didn't want to send it in for replacement when it became sick. One wrote, 'I can't imagine not having him any longer. He's my BABY!!' And three respondents, when asked to provide names and ages of family members in the household, provided demographic information for their Roombas.

We can't avoid sympathizing with robots. A European study found that subjects' mirror neurons charged up whether they watched a human hand or a robotic claw perform various actions, such as picking up a cocktail glass. 'Now we know', the researchers wrote, 'that our [mirror neuron system] may be part of the reason why, when in *Star Wars*, C3PO taps R2D2 on the head in a moment of mortal danger, we cannot help but attribute them human feelings and intentions, even if their physical aspect and kinematics are far from human.'

The behaviour of computers does not even need to be embodied in physical movement to appear agentic, as anyone who has growled at his laptop for losing a file will tell you. Clifford Nass, a sociologist at Stanford University, in California, has found that human-computer interaction closely mirrors human-human interaction; it automatically triggers certain social expectations. For example, people implicitly apply gender stereotypes about friendliness and competence to computers based on the gender of the computer's speaking voice, despite denying any difference between male-voiced and female-voiced computers. People also attempt to be polite to computers: after working with one computer, they'll give higher ratings of that computer when answering a questionnaire on that same machine than on an identical one or on a piece of paper. They'll also reciprocate assistance: if a computer provides useful versus useless search results, subjects are more likely to help the computer create a colour palette. And users will reciprocate self-disclosure: they offer more personal information in response to the prompt 'There are times when this computer crashes for reasons that are not apparent to its user. . . . What have you done in your life that you feel most guilty about?' than to the question alone. You want to share something of yourself in response to the silicon's vulnerability. It's like a little heart-to-hard-drive.

Designers of robots, gadgets, and computer interfaces are striving to make the interfaces 'invisible', so that instead of dealing with a machine, you're collaborating with a partner. As the technology around us becomes more responsive and interactive, 'it will make sense to have a degree of magical thinking just to be able to deal with the devices,' Erik Davis, the author of *TechGnosis: Myth, Magic, and Mysticism in the Age of Information*, told me. One of the dangers of applying social expectations to silicon is that your new e-friend will at some point let you down; we haven't advanced past the 'artificial' in AI.

God

The term *anthropomorphism* comes from the Greek philosopher Xenophanes, who 2500 years ago described the way we see gods in our own image: 'Ethiopians say that their gods are snub-nosed and black;

Thracians that they are blue-eyed and red-haired.' He even speculated that if horses and oxen and lions could paint, they would depict their gods to look like themselves. Rumour has it Xenophanes had a killer 'white people pray like *this* and horses pray like *this*' stand-up bit.

Stewart Guthrie notes in *Faces in the Clouds* that around the world and across eras, gods have existed on a spectrum with humans, eating and drinking and dying and making love and war with people. Contemporary Western religions make more of an effort to separate God from the riff-raff, giving him such otherworldly qualities as omnipotence, omniscience, omnipresence, and indifference to reality television. But our intuitions don't always stick to scripture. The psychologists Justin Barrett and Frank Keil told various fictional stories about divine intervention to a set of study participants, then asked the subjects to retell the stories in their own words. The subjects had already affirmed that God knows everything, is everywhere, can read minds, and can multitask, but their story retellings recast the Almighty as basically a superhero with limited bandwidth. Subjects automatically reshaped the material in their heads so that God had to, say, finish answering one prayer before attending to another, and could not even hear a pair of birds above the noise of a jet engine.

People also attempt to communicate with God as if he were, well, a *he* (or a *she*), rather than an *it*. The sociologist Christopher Badcock wrote that we believe in 'divine agents who can be influenced in mentalistic ways analogous to those in which ordinary humans can be: through supplication (prayer), flattery (praise), generosity (sacrifice), apology (confession), restitution (penance), visitation (pilgrimage), and lobbying (intercession via saints, angels, or other deities).'

And we read God's mind as egocentrically as we do other people's – actually more so. Nicholas Epley and collaborators asked subjects about their own beliefs, God's beliefs, and other people's beliefs on several issues including abortion, same-sex marriage, and the war in Iraq. God's views correlated with their own views more strongly than others' did. To show that God's purported attitudes echo subjects' attitudes and not vice versa, Epley manipulated participants' takes on the death penalty and found that God's opinion budged too. Finally, the researchers found that the parts of the brain recruited when thinking about God's beliefs resemble the areas used in pondering one's own beliefs more closely than the areas used in pondering other people's beliefs. Epley suggests three potential reasons we're more egocentric with God's mind than with our human peers. First, we don't have a lot of reliable background information on God, so we just use ourselves as a model. Second, disagreement with God would be threatening. And third, we assume we believe what is true, and God believes what is true, so God should believe what we believe.

In light of Epley's research, the fact that half of all Americans consult God for advice daily should give us pause. Praying for guidance is about as productive as asking your imaginary friend where you left your keys – he doesn't know any more than you do. It's also more dangerous, because a divine stamp of approval on your own fallible hunches might send you on a wild goose chase with extra vigour. As Bob Dylan sang in 1963, cynically reprising a go-to justification for war, 'That land that I live in has God on its side.'

But anthropomorphism is not only unavoidable in religion; as many theologians have noted, it's the *foundation* of religion. If God were merely an impersonal phenomenon, to whom would one pray? Some people try to avoid reducing God to a humanlike agent by denying him any definable qualities – by saying he is everything and nothing – but such an entity is incomprehensible and thus meaningless. According to the 19th-century German philosopher Ludwig Feuerbach, 'The denial of determinate, positive predicates concerning the divine nature is nothing else than a denial of religion . . . it is simply a subtle, disguised atheism'.

So we perceive humanlike qualities in God, but he's still somehow different. I suggested earlier that one could think of mental presence as lying on a spectrum rather than existing as a discrete quality, with animals somewhere between humans and rocks. But we actually think about mind in terms of two distinct dimensions, Heather Gray, Kurt Gray, and Daniel Wegner have found; there's *agency* – the capacity for self-control and planning – and *experience* – the capacity for states such as pain, pleasure, and fear. We attribute both agency and experience to adults, experience but not agency to babies (they're sensitive but not very useful), and agency but not experience to God (he wields great power but is fairly immune to anything that might befall him). Robots are low on experience and at the midpoint on agency, and animals and foetuses are low on agency and with some amount of experience. So as the researchers note, having a mind, in others' eyes, is not simply a matter of presence, or even degree of presence, but of type. Is the mind a doer, a feeler, or both? God is a doer.

The significance of treating God as a doer, in particular a doer who shares human concerns and motivations, will become apparent in the next chapter as we discuss destiny.

Detecting Agents

Egocentrism can explain the anthropomorphism of nonhuman agents, but what makes us treat inanimate objects as agents in the first place? What features bring something into the realm of lifelike creature and make it worth trying to psychoanalyse? Let's look at three triggers: behavioural, morphological, and moral.

Behaviour

The most obvious sign of agency is movement. In a seminal study published in 1944, Fritz Heider and Marianne Simmel showed Smith College students a simple two-and-a-half-minute black-and-white animation featuring a large triangle, a small triangle, and a circle, and asked them to write down what they saw. Of thirty-four subjects, thirty-three spontaneously described the shapes in animate terms, usually as people.

Nineteen subjects connected the series of events into a narrative. One wrote, for example, 'Lovers in the two-dimensional world, no doubt; little triangle number-two and sweet circle . . . we regret that our actual knowledge of what went on at this particular moment is slightly hazy, I believe we didn't get the exact conversation.'

Other researchers have stripped down their animations to locate the minimum and essential movement cues for seeming agentlike, in some cases showing subjects a simple dot moving across a screen. A change in behaviour in the absence of obvious outside causes – starting from a stop, or shifting speed and direction without bouncing off anything – often appears lifelike. Movement contingent upon the environment or another agent, such as avoiding an obstacle or chasing another entity, works especially well: living things react to their world and interact with other living things.

Sometimes we *imagine* motion. As Guthrie wrote, 'Banging our eye on an unexpected door edge, we kick the door, because we see it in that moment, automatically and unconsciously, as a malefactor. *It appears to have struck us.*' We weren't predicting the door in our path so its mere presence feels like a deliberate act, and we reflexively go into revenge mode. This is why when people trip on a root, they say, 'Where did that come from?' An ex-girlfriend still teases me about the time I stomped on an egg to punish it for jumping out of my hand onto the kitchen floor. *I* certainly didn't mean for it to drop.

When we anthropomorphize triangles and doors and other objects, is it just an exercise in metaphor, or is it a bubbling up of instinctual mind detection – our impulse to see intentionality in behaviour? Well, babies are notoriously weak with metaphors, but they get in on the anthropomorphism action as soon as they start thinking about minds in people. Infants as young as twelve months prefer watching a shape approach another shape that 'helped' it earlier to watching a shape that was mean to it. They appear to see the shapes as social actors. (Of course in studying anthropomorphism *in* babies, researchers make efforts to avoid anthropomorphism *of* babies; they try to rule out simpler mental processes not involving theory of mind that could explain the infants' behaviour.)

Neuroimaging also points to genuine mind-attribution in anthropomorphism. Watching animated shapes move in a goal-directed way engages parts of the brain used for detecting biological motion, such as the superior temporal sulcus, as well as areas employed in judging others' mental states, such as the medial prefrontal cortex. The woman quoted in Heider and Simmel's 1944 study was not just having a gas; some part of her really saw the triangle and circle as lovers.

Morphology

In 1996, a short report titled 'The Case of the Haunted Scrotum' appeared as the last item in the *Journal of the Royal Society of Medicine*. A forty-five-year-old man had visited J. R. Harding, a radiologist, for examination of an undescended right testis. Harding didn't find it with an ultrasound, so he performed a CT scan and saw something odd on the left side: what appeared to be 'a screaming ghostlike apparition'. He was kind enough to include an image in the report. And as for the right testis? 'None was found', he wrote. 'If you were a right testis, would you want to share the scrotum with that?'

We see faces everywhere. In 2004, a woman sold a cheese toastie that appeared to have the Virgin Mary's visage on it for $28,000 (about £15,000). In 1976, NASA's Viking 1 spacecraft sent back an image of a rock formation on Mars that distinctly resembled a human mug. And a famous photograph from 2001 shows a demon apparently grinning in the smoke billowing from the World Trade Center.

We process these images much the same way we process real faces. One study measured brain activity in subjects as they saw photos of faces; photos of everyday objects vaguely resembling faces, such as an electrical socket; and photos of other objects. Both the faces and the facelike objects activated an area of the fusiform gyrus called the fusiform face area (FFA) that responds strongly to faces. And the sockets and such didn't tickle the FFA as a result of a considered top-down reinterpretation; they triggered facial processing systems automatically, within 165 milliseconds of appearing. Which means we see faces in clouds before we can even say 'cloud'.

Seeing faces or other elements of the human form in random noise is not in itself magical thinking; it's just an example of *pareidolia*, the finding of patterns in vague sensory stimuli. But physical similarity to humans can trigger or amplify the attribution of human mental qualities to these nonhumans, which *is* magical thinking. The brain reacts to the emotions displayed by robotic faces within about 100 milliseconds, which is as long as it takes to recognize emotion in human faces. The brain will even react to the sentiments expressed by simple smiley faces in about that time. And in one study, twelve-month-olds followed the 'gaze' of a brown furry lump with felt patches for eyes and a nose. Those felt patches were enough to engage the baby. Eyes are particularly important for expressing animacy, as we read so much into what's behind them. As Dorothy Marette, the psychologist at the Veterans Affairs Medical Center, said, 'When Paro looks over at you with those eyes, you can't help but say, "Hello".'

Robot designers must be careful not to make their creations look *too* much like a human, however, unless they can get every detail right. Otherwise they run the risk of entering the 'uncanny valley', that terrain situated between stylized and spot-on depictions of life-forms where things are just off enough to give you the willies. (Recall the human characters in the animated movies *The Polar Express* and *Final Fantasy*.) Takanori Shibata modelled Paro after a seal because true animal realism was out of reach. He designed dog and cat prototypes, but people preferred interacting with a creature whose behaviour didn't clash with their expectations. They knew too much about how dogs and cats *should* act, but it turns out that the limitations of the robot's capabilities fall just behind the limitations of people's knowledge of baby harp seals.

We read into the face on the seal, and we also read into the faces of babies and foetuses. In 2010 a British couple received an ultrasound image in which their four-month-old fetus appeared to be smiling. The *Daily Mail* writer who reported the story speculated, contra the experts she interviewed: 'The scan implies that a baby can experience feelings such as happiness and pain much earlier in its development than previously thought.' As a writer for *Gawker* put it, 'The abortion debate has devolved to the zygote version of a LOLCat', referring to those funny cat pictures on the Internet with ungrammatical captions. She added that the foetus 'is 17 weeks old, and much like a LOLCat, people keep trying to put words in its mouth. They ask, *What does LOLFetus have to say about abortion politics?*' In reality, the apparent smile is probably no more than an example of pareidolia. The bulbous head with a creepy grin actually looks a lot like the haunted scrotum found by J. R. Harding. Twins, separated at conception?

A friend of mine utilizes the evocative power of faces to avoid losing her stuff. She goes to festivals a lot and has lost bags and other items but hasn't lost her current bag because she's painted a face on it. And she used to lose water bottles frequently, but not the current one, with a face on it. The faces make her take better care of her things. 'I don't leave stuff behind', she says, 'because if I start to walk away, I'll think, "I'm forgetting someone."'

Morality

In ancient Greece, a statue of the athlete Nikon was once toppled off its pedestal by Nikon's enemies, crushing one of the perpetrators. The statue was brought to trial and sentenced to be cast into the sea. When Ivan the Terrible's youngest son died in the Russian town of Uglich in 1591, the town bell was rung in celebration. It was banished to Siberia for its political offence and not fully pardoned until 1892. In the 1940s a young bell ringer at the Mexico City Metropolitan Cathedral died when the bell struck him in the head. The bell was silenced as punishment, until its pardoning in 2000.

Throughout history and around the world, animals and inanimate objects have been formally tried, convicted, and punished for their involvement in negative events. It's the old-school equivalent of suing McDonald's when you spill hot coffee on yourself. Something bad happens and you need someone, or something, to blame.

But why does there need to be blame? Can't bad things just happen? For every moral action involving help or harm, there is a doer – a *moral agent* – and a receiver – a *moral patient*. Kurt Gray and Daniel Wegner argue that this dyadic template of morality is so prominent in our thinking that whenever we see someone harmed we automatically apply the template, casting that person as a moral patient and looking for someone to play the role of agent. If no human scapegoat is available, we cast a real goat, or a statue, or a bell. If you think as a modern society we're above ceremonial execution of inanimate objects, recall from chapter 2 what happened to the baseball that eluded the embrace of the Chicago Cubs player Moises Alou in the 2003 championship playoffs: it was detonated before a television audience.

Agents Everywhere

Autonomous behaviour, the appearance of a face, and participation in harmful events all seem like reasonable signs of agency in certain circumstances, but what doesn't seem reasonable is the willy-nilly treatment of anything that displays any of those signs as having a mind. Shouldn't we be more discriminating? Wouldn't we be better off if we personified less promiscuously?

It turns out that sluttiness pays.

Agents typically play the starring roles in our environment. During evolution, we had to pay attention to predators and prey so as not to miss a meal or become one. And still today other humans offer cooperation and competition as affiliates or adversaries; whatever the people around us are up to, we want to know. So as long as we're unable to identify instantly and with complete accuracy what's an agent and what's not, it's better to go on the assumption that something *is* an agent until proved otherwise. Our eager agency detection system is a demonstration of error management theory (discussed in chapter 3): if one type of error (a false alarm) harms us less than the alternative error (a miss), we'll increase our false positives, even at the expense of more overall errors. As Stewart Guthrie wrote in *Faces in the Clouds*, 'It is better for a hiker to mistake a boulder for a bear than to mistake a bear for a boulder.'

Animals have the same bias. If you've ever seen a scarecrow at work, heard your dog defend you from an intruding wheelie bin in the middle of the night, or played with a cat and a laser pointer, you know that liberal animacy detection is not restricted to humans.

Fear further biases our judgement toward seeing agents – the bump-in-the-night phenomenon. EWC, together with Scott Akalis, found that people are more likely to interpret abstract sketches as faces after watching three minutes of *Silence of the Lambs*.

We're also better off overestimating the probability of someone witnessing us harming someone else, because a witness or his friends can harm us. Some psychologists have suggested that fear of third-party condemnation has aided the evolution of religion. The predominant view among psychologists of religion is that belief in gods and spirits is a side effect of basic cognitive mechanisms – such as dualism (the belief that spirits can exist independently of bodies, explained in the last chapter) and a tendency to infer a designer behind significant objects and events (explained in the next chapter) – combined with catchy cultural traditions that build on these cognitive inclinations. But Jesse Bering, Dominic Johnson, and a few other researchers argue that the idea of a supernatural judge became so useful to our ancestors that those who believed in it most strongly had a survival advantage over their peers, and that evolution then specifically promoted a religion-friendly disposition. The habits of mind underlying religious belief, then, would be a bit like feathers, which evolved for one reason (insulation) but then became handy for something else (flight) and eventually specialized to assist in their new function. We have a dedicated system in our brain for thinking about God, they argue.

Belief in God is useful because God enforces morality. If you act selfishly and get away with it, you obtain an advantage. But if someone catches you, you're perhaps disproportionately disadvantaged – your reputation is ruined, you may be expelled from the group, and you may even be killed. So, according to error management theory, it would be better to overestimate the likelihood of an anthropomorphized agent watching you. And if you believe in invisible spirits who are interested in the affairs of humans – in anthropomorphic gods and ghosts – then you always have a witness over your shoulder. Indeed, the researchers Azim Shariff and Ara Norenzayan found that a subtle reminder of God encourages subjects to act generously in an anonymous economics game.

But you needn't believe in God to act morally. My conscience guides me without any (explicit) fear of magical punishment. Psychologists have a few explanations for anonymous altruism – prosocial behaviour that doesn't benefit oneself, one's kin, or one's reputation. One possibility is that we do good because we overestimate the chances of another *human* finding out about it. The piece of wisdom expressed by the author H. Jackson Brown Jr's mother that 'Our character is what we do when we think no one is looking' may be based on a false premise – that we ever think no one is looking. For a counterpoint, consider what the satirist H. L. Mencken wrote (albeit somewhat sardonically): 'Conscience is the inner voice that warns us somebody may be looking.' In line with this suggestion, Shariff and Norenzayan found that while reminders of God didn't significantly increase the anonymous generosity of atheists, subtle exposure to words like 'police' and 'jury' did; just the idea of recrimination heightened their paranoia and their prosociality. Other research shows that after pouring coffee people place more money in a volunteer donation box when a simple photo of human eyes is displayed nearby.

Another explanation for kindness without recompense is that we internalize social norms. Yet another is that ensuring people get what they deserve, whether help or harm, affirms our belief that the world is just. And a fourth proposes that altruism emerges from our capacity for empathy and simulation; we don't want to harm others because we feel their pain. There are many ways to be good without God.

The Need for Control

Adidas introduced a new eight-panel ball for the 2010 World Cup. The ball was named the Jabulani, Zulu for 'celebrate', but not everyone celebrated the ball. Some critics took to calling the ball 'Jumanji', after the book in which wild animals come to life from a game.

'Obviously, it's quite unpredictable, the way the ball flies', the Australia keeper Mark Schwarzer told the BBC. 'Sometimes the ball has a genuine flight, and other times it has a mind of its own.'

'It's very weird', the Brazil striker Luis Fabiano told the Associated Press. 'All of a sudden it changes trajectory on you. It's like it doesn't want to be kicked. . . . You are going to kick it, and it moves out of the way. I think it's supernatural.'

In play, ball control is a must, and on elite fields it's a given. So when a ball's behaviour defies expectations, players turn to animacy: the football must have a mind of its own.

People will similarly shout at their cars or their computers when these machines malfunction – when they disobey orders.

We react to uncertainty by forming hypotheses, and hypotheses that generate a lot of predictions usually win out over others; they're more useful. Betting on the presence of an agent allows for all kinds of predictions, because we understand minds and are familiar with many of the goals they might have. Minds make flailing bags of skin comprehensible. In Guthrie's words, when hikers bet on the presence of a bear (versus a boulder) and are right, 'the jackpot is whatever they know about bears'. For instance, that bears sometimes like to eat hikers.

Feeling especially mystified or anxious will increase our *effectance motivation*, a term coined in 1959 by the Harvard psychologist Robert White to describe our persistent urge towards competence – towards an ability to understand, predict, and control our environment. As we saw in chapter 3, when people feel out of control they'll attempt to make up for it with superstitious rituals. And thinkers dating back at least to Hume have suggested that a desire for competence pushes us towards explaining the world using anthropomorphism. According to Freud, 'If everywhere in nature there are Beings around us of a kind that we know in our own society, then we can breathe freely . . . we can try to adjure them, to appease them, to bribe them, and, by so influencing them, we may rob them of a part of their power.'

In an experiment EWC conducted with Scott Akalis, subjects completed a measure of their desire for control, then watched a video of a small, quick, unpredictable dog and a large, plodding one. Overall, subjects said the unpredictable dog had more personality and conscious will than the larger one, and this difference was amplified in subjects with a high desire for control.

Then EWC ran a series of experiments with Carey Morewedge, George Manteleone, and Jia-Hong Gao. In the first, they found that people who reported more problems with their computers were more likely to say their computers behaved as if they had their own beliefs and desires. In the second, subjects read about thirty different robotic gadgets, described as either predictable or unpredictable. For example, Clocky the alarm clock can be programmed to either run away from you or jump on you when you hit SNOOZE (predictable), but Pillow Mate will decide on its own whether to hug you or curl into a ball when squeezed (unpredictable). People attributed more intentions and emotions to the unpredictable devices.

Maybe you're thinking at this point that people aren't anthropomorphizing to reduce unpredictability; they just associate unpredictability with humans. Absolutely, seeing variation in an entity's routine can be reminiscent of people, who typically don't act like assembly-line robots, but acting completely erratically doesn't seem intentional either; it's a sign that one has *lost* one's mind. In any case, subjects told the experimenters that predictable behaviour is more typical of humans than unpredictable behaviour is. Pillow Mate was not just a reminder of a familiarly fickle bedmate; subjects were trying to understand it.

Rating unpredictable gadgets on their level of agency also produced more activation in parts of the brain used for mind reading than rating the predictable gadgets did. 'Perceiving an agent as having a mind of its own may not be mere metaphor', the authors wrote.

If competence motivation increases anthropomorphism, then anthropomorphism should increase a sense of competence. Subjects watched clips of a dog, a robot, Clocky, and animated shapes and were told to get inside the heads of two of them, and to treat the other two as a behaviourist would, noting only objective behaviours. Afterwards, they said they felt more able to predict the future behaviour of the entities they psychoanalysed.

But does magical thinking of the anthropomorphic variety provide *actual* competence or just the feeling of competence? The philosopher Daniel Dennett refers to the inference of an organizing force – a mind – behind behaviour as 'taking the intentional stance'. Other options include taking the 'physical stance' – applying what one knows about natural laws – and taking the 'design stance' – considering the intended purpose and functionality of an entity. Each stance has its advantages and disadvantages. You could take the intentional stance and explain a chess-playing computer's sitting still on a table by saying it 'wants' to remain there, rather than taking the physical stance and appealing to gravity and inertia, but you would sound silly and, worse, you might go on to make mistaken predictions about its behaviour, such as the expectation that a good pep talk could convince it to move about. On the other hand, taking a physical stance won't help you beat it at chess; you want to know which strategies it's considering, not how many electrons it's sending through each circuit. (The design stance tells you how to turn the thing on and start the program.)

So the heuristic of taking the intentional stance with a computer could be considered productive anthropomorphism, at least until you try to guilt it into not taking your pawn. And gardeners talk about 'tricking' plants into blooming at certain times. But there aren't many other cases where anthropomorphism allows for actionable predictions. Taking the intentional stance with an unpredictable football certainly doesn't provide any traction on the situation.

Loneliness

'The bad thing about being in here is you don't get to really have a relationship with anybody.' Pierre Carter is describing to me his life as a resident at the Veterans Affairs Medical Center in Washington, DC. Carter, who suffers from post-traumatic stress disorder, is a large man in a wheelchair wearing reading glasses, a baseball cap with a military insignia, and earrings, including a peace sign dangling from his right ear. The hospital's Paro robot, which he calls Fluffy, rests on his belly looking up at him, appearing to suckle on the shock of white goatee extending from his chin. The shared colour of their fur contrasts with the black biker gloves Carter uses to gently stroke the seal. 'Being in here, it's lonely. It's very, very lonely', he says. 'The animal, it's almost like it's a real individual, and it gives you back what you give it. It kind of keeps you company. . . . When she looks right at you, goodness gracious.'

Humans have a fundamental need to belong, and over the years people have suggested that we anthropomorphize to populate the world with allies. A Greek folklorist argued that ancient navigators personified mountains and rocks in response to solitude. The screenwriter for *Cast Away* gave Tom Hanks's character a companion in a volleyball named Wilson. And the proverbial crazy cat lady always lives alone.

EWC and Scott Akalis brought back Clocky to study how loneliness would affect people's interpretations of an alarm clock that runs away from you. Subjects rated Clocky and three other gadgets on the degree to which they had emotions and free will, then noted how often they themselves felt isolated from other people. The most lonely subjects gave the most anthropomorphic ratings. In another study, inducing loneliness by showing a clip of *Cast Away* increased subjects' anthropomorphism of their pets.

A third study followed from the hunch that isolation makes us not only anthropomorphize agents around us but also believe more strongly in commonly anthropomorphized religious agents. (Previous research had shown that both singles and people who'd grown up with cold mothers tended to turn to God for accompaniment.) Subjects made to feel like loners in the lab reported greater credence in several supernatural agents – ghosts, angels, God, and the Devil – as well as acts perpetrated by these agents – miracles and curses.

Does magical thinking provide the same benefits as real social connection? An inflatable doll does not a real girlfriend make, but pets increase survival rate, reduce blood pressure, and ameliorate depression in their caretakers. And Shibata and collaborators have conducted several studies on Paro's therapeutic value over the past several years. The robot improves people's moods and makes them more active and communicative. After leaving Paro in a public area in a nursing home for five weeks, researchers found that residents had stronger social ties to each other, and urine tests revealed biomarkers for reduced stress. And in a study of fourteen patients with dementia, an EEG revealed improved cortical functioning in half the subjects after a single twenty-minute interaction with our favorite robotic seal.

'A lot of veterans, particularly with PTSD, don't trust very easily', Dorothy Marette at the VA Medical Center says. 'They don't disclose very easily, they don't connect very easily. And it's a safe way to start. That unconditional love you get from an animal, you don't worry about being judged and what they're going to think about you.'

'It makes you bring out your deepest feelings, and they don't discriminate', Carter says of the robot. 'Whatever you give it, it gives you back five hundred percent. It lets people know that you don't really have to be special, but it makes you feel special.' He directed some half whispers at Fluffy. ('How you doin', how you doin', how you doin', all right, all right, all right.')

'It makes everybody feel like a king.'

Treating Humans as People

So we're adept at seeing minds because seeing minds is a useful skill. It allows us to predict the behaviour of predators, prey, and people, and to interact with them. We look for minds especially when we have a need for control or when we feel lonely. And our mind reading continually spills over to entities that don't actually have minds.

The consequences of this spillover include the relationships we build with animals, the trust we accord technology, the blame we direct at gods, the protection we provide our environment, the desire we have for cute consumer products, and the metaphors we use to understand a range of phenomena in fields such as science, economics, and politics (selfish genes, bullish markets, the war on drugs). But anthropomorphism may have its greatest implications when directed at people.

Despite research showing people act largely on autopilot, the psychologist Evelyn Rosset has argued that by default we see everything anyone does as intentional. Seeing an unintentional act as intentional, I argue, constitutes momentary anthropomorphizing. You're looking at a human, a biological creature, performing some act, perhaps inattentively, and imputing full awareness of the action and its repercussions. Say your son tracked mud inside the house. Was he doing it to annoy you? Maybe he was busy texting and didn't even notice his shoes were dirty. Humans can't be aware of everything they do at all moments.

In one study, Rosset gave subjects either 2.4 or 5 seconds to judge whether various acts were done on purpose. For actions always or typically done accidentally ('She kicked her dog'), time pressure increased ratings of intentionality. In another study, reducing participants' inhibition by getting them drunk also increased their rate of finger-pointing, a finding that explains why harmless jostles at the pub so easily escalate to inebriated brawls. I often say that life would go a lot more smoothly if we just assumed people were idiots instead of assholes. (But even smoother if we gave people the benefit of the doubt all around.)

Beyond everyday misunderstandings, we endanger ourselves and those we should be looking after when we overestimate the level of moral and criminal responsibility in children and the mentally ill or disabled. As discussed in the section on free will in chapter 4, we shouldn't jump to call these offenders – or anyone – evil.

Anthropomorphism of damaged or elderly brains also complicates the difficult issues of life support and euthanasia. Is the person we know still there inside that head, or are we projecting?

And encouraging the personification of foetuses has been a primary strategy in anti-abortion campaigns. In several US states, women opting for an abortion are now legally required to undergo an ultrasound and be offered the chance to look at the image.

It's worth remembering – whether you're navigating rush-hour traffic or debating public policy – that even humans can be mindless sometimes.

|||||||||||||||

In healthy, nonautistic adults, mind perception is easily triggered, but it still needs triggering. The opposite of anthropomorphism is another constant in human history: dehumanization. We often treat social outsiders, the elderly, enemies, children, and anyone we have no particular motivation to engage with as less than equal. Overattributing mind through magical thinking may bring to life otherwise inert elements of the world, but the danger of *under*attributing mind lies in robbing your fellow humans of their personhood. They become animals or objects, fit for disregard, mistreatment, even elimination. EWC note irony in the fact that 'the same rights conferred to animals, plants, or rivers through anthropomorphism can be denied to people'.

It's a denial of rights that sometimes occurs in settings like the Veterans Affairs hospital. Although anthropomorphism of robots (and animals) can aid in therapy, as demonstrated by the studies of pets and Paro, 'a lot of people are afraid that Paro will serve as an inanimate object that goes between people and directs the patient's response to a robot instead of to a real person', Dorothy Marette says. According to this argument, by leaving the patient to anthropomorphize Paro, we're dehumanizing the patient.

Marette uses Paro largely as an icebreaker, however. 'I've found that maybe the patients wouldn't have spoken to me too much, but they start off feeling freer to interact with the robot,' she says, 'and then they look up and start talking to *me*.'

That's when the real magic happens.

7

||||||||||||||||||||||||

Everything Happens for a Reason

You've Got a Date with Destiny

When Cindy Bennett was thirty-five, she held a wake for a daughter who was never born. A daughter who was not only never born, but never conceived.

Bennett used to stick her stomach out in front of a mirror to see how she would look pregnant. She watched her two older sisters give birth and knew that she would do it better than they did. And she looked forward to raising a daughter, a girl who would look like a combination of her and her husband, Randy, and would be bright and funny and assertive and love her mother dearly.

But when she and Randy went into reproduction mode, they had problems. Whatever was supposed to happen with the birds and the bees just wasn't happening. So they sought help.

The Bennetts started infertility treatments, but one of Cindy's doctors felt something was particularly not right during one of her exams and decided to take an ultrasound. She found a large cyst on Cindy's left ovary. Although the doctor had little reason to suspect cancer, she decided to remove the cyst whole and perform a biopsy rather than suction it out laparoscopically. Inside, she found a tumour. Cindy had stage 1, grade 3 ovarian cancer.

Cindy asked her surgeon to save as much of her reproductive system as he could, but he told her he couldn't make any promises. The cancer was aggressive, 'so they kind of had a fire sale', Cindy told me. 'They took whatever they could. When I woke up they let me know. That's kind of when it hits you. The prognosis is good, you know you're going to live, but you've got a whole new set of issues to deal with, which is that you're now infertile.'

ıııııııııııı

A few years ago, Miriam Klevan, a researcher at Northwestern University, in Chicago, interviewed nineteen couples about their experiences with infertility. Most of them said it was the worst thing that had ever happened to them. 'It brings them into real contact with mortality, just being stymied in one of the deepest ways possible, and feeling extremely inadequate', she told me. 'The word *desolate* came up over and over.' Klevan and her husband had dealt with infertility themselves. 'I felt like I was dying. And I think this is how a lot of people feel', she said. 'From an evolution standpoint, of course it makes sense: I was facing the death of the line.' Many couples also feel a deep sense of failure. 'This is something teenage girls do all the time, and we can't.'

'I was angry. I was really angry', Cindy says. 'And I hated anybody who was pregnant.' Her oncologist worked out of an obstetrician's office, so the waiting room was full of pregnant women, and the surgery was full of baby pictures. 'So, yeah, that sucked, and I hated everybody. Any time something is taken away from you that just feels natural to everybody else, it's difficult.'

Cindy also felt she had let Randy down. 'One thing we worked on in couples therapy was my not being able to give something to my husband', she told me as I sat with her and Randy at their dinner table in suburban Virginia. 'It took me a long time to feel really secure that I hadn't disappointed him, the fact that I couldn't get pregnant. That was a tough one. I didn't believe him for a long time that he wasn't sorry.'

The brush with death, the infertility, the marital strain: Cindy was dealing with challenges her sisters hadn't had to face. She struggled to find a reason for it all. 'You don't know how you're going to deal with it. And you can't understand why it happened to you.'

The Missing Link

In Ernest Hemingway's short story 'The Snows of Kilimanjaro', a writer named Harry goes on safari with his lover, Helen. A thorn scratches his knee as he photographs a herd of waterbuck, and the wound becomes infected. The story opens with Harry dying on a cot in the shade of a tree as birds circle above.

'I don't see why that had to happen to your leg', Helen says. 'What have we done to have that happen to us?'

Harry offers several reasons. 'I suppose what I did was to forget to put iodine on it when I first scratched it. . . . If we would have hired a good mechanic instead of a half baked kikuyu driver . . .' After each explanation, Helen responds with 'I didn't mean that.'

There's a miscommunication. Harry is explaining *how* they ended up in this situation. Helen wants to know *why*.

When asking *why* (or *why me*), we're demanding some reason beyond 'this happened and then this happened' – some answer that will give the event a higher meaning, that will put it into context and maybe even excuse it. We're hoping that it's all part of someone's or something's plan. Asking *why* taps into the same sort of magical thinking as saying that something was *meant* to happen, or that everything happens *for a reason* (an adage as aggravating to some as it is analgesic to others). We're picturing a universe that cares about human affairs, particularly our own.

Magical reasoning doesn't always replace mechanical reasoning. To many, supernaturalism doesn't supercede naturalism but works through it. Divine intervention has both a *why* and a *how*.

The famed British anthropologist E. E. Evans-Pritchard lived with a Zande tribe in the southern Sudan in the late 1920s and found witchcraft to be part of their everyday lives. The Azande blame all kinds of misfortune on witchcraft, from poor crop yield to marital trouble to stubbing one's toe. Blaming a witch comes as naturally to them as profanity does to us. But they don't invoke magic to explain why crops fail in general – only why one's own crops failed. (*Why me?*) And even then, they acknowledge the natural causes. Evans-Pritchard noted that occasionally an old granary would suffer termite damage and collapse. And sometimes it would injure people sitting under it for shade. The Azande understand termites, and they understand the need for shade. 'That it should collapse is easily intelligible,' Evans-Pritchard wrote, 'but why should it have collapsed at the particular moment when these particular people were sitting beneath it? . . . Zande philosophy can supply the missing link.'

People see both witchcraft and God's will as intervening through natural means. Research by the psychologist Michael Lupfer shows that for important life events, American university students tend to credit God in tandem with personal and environmental causes. The idea of a God of the gaps, often used pejoratively to describe the use of religion to explain only what science hasn't covered yet (thus condemning one's god to ever-decreasing utility), may not be a fair representation of religion for many people. Filling a gap with science still leaves room for God.

At least it does in the cosmologies of the faithful. It should be pointed out that if cold natural causes are sufficient to explain a phenomenon, there actually isn't any place for divine intervention. If you can predict a termite's path perfectly with unvarying physical laws, a little nudge from above would necessarily violate those laws. Which leaves God the freedom only to assign initial conditions to the universe, to invent the laws and get things moving in the first place. But a god sitting impotently on the bench for the last fourteen billion years or so isn't what most people think of when they think of God. They see him as an active meddler.

So integrating God (or witchcraft) with physics doesn't help us accurately predict the behaviour of the world, and in fact it gets in the way of accurate predictions. But, as we'll see, what magical thinking can offer is an increased *sense* of predictability and, perhaps more important, a sense of *purpose*.

The End Is in Sight

In the last chapter we broached the topic of religion. We saw how our bias towards treating entities as people leads to the anthropomorphism of God. But what that chapter didn't address is how we come to infer the existence of a god in the first place. The strongest triggers of anthropomorphism – how something looks and how it moves – don't apply to entities you can't see. We can deduce God's existence only indirectly, by observing the purported effects of his will.

We see the effects of his – or someone's – will everywhere. Nearly every culture has a creation myth describing the design and genesis of the world. Scholars have written for thousands of years about the uncanny amount of order and beauty in our environment. Cicero, the ancient Roman philosopher, offered the first recorded watchmaker analogy: 'When we see some example of a mechanism, such as a globe or a clock or some such device, do we doubt that it is the creation of a conscious intelligence? So when we see the movement of the heavenly bodies . . . how can we doubt that these too are not only the works of reason but of a reason which is perfect and divine?'

Arguing that because something has a function it was intentionally created for the sake of that function is what's called *teleological* reasoning. (*Telos* is Greek for 'end' or 'purpose'.) Teleological reasoning is an appropriate response when looking at a watch – it tells time and was indeed created for the purpose of telling time. But some things that look designed result from natural events. 'We are entirely accustomed to the idea that complex elegance is an indicator of premeditated, crafted design', Richard Dawkins wrote in *The Blind Watchmaker*. 'This is probably the most powerful reason for the belief, held by the vast majority of people that have ever lived, in some kind of supernatural deity.'

Perhaps the most stunning piece of apparent craftsmanship, one that fools even some scientists, is the grandeur of organic life, with the human species as its crown jewel. In 2010, a century and a half after Charles Darwin published *On the Origin of Species*, 40 percent of American adults polled by Gallup said God created humans in their present form. Another 38 percent said we evolved, but with God guiding. Only 16 percent said we evolved with no magical intervention. 'It is almost as if the human brain were specifically designed to misunderstand Darwinism, and to find it hard to believe', Dawkins wrote (with presumably intentional irony).

The psychologist Jean Piaget argued that all children are 'artificialists'; he thought that for the first several years of life we can't reason in terms of natural physical causes, and so we believe everything in the world is created by humans. It's since been shown that even babies understand simple physics, but Deborah Kelemen at Boston University has confirmed in multiple studies that children are at least biased to think objects were created for a purpose. About three in four preschool children she tested assigned functions to natural objects: clouds are 'for rain'; mountains exist 'to climb'; lions were made 'to go in the zoo.' Even ten-year-olds prefer teleological explanations to physical ones: rocks are pointy so animals can scratch themselves on them, not because little bits of stuff pile up. Kelemen calls children 'intuitive theists'.

While adults may learn more accurate explanations for objects' origins, we don't outgrow our preference for teleology; we merely constrain it. One study gave grown-ups either three or five seconds to judge whether statements presented to them were correct. Those on the shorter clock supported teleological arguments ('the sun radiates heat because warmth nurtures life'; 'molecules fuse in order to create matter') more strongly than those with more time to think about their answers. We assume design by default.

|||||||||||||||||

Sometimes the surprisingly elegant way the pieces of our own lives fit together can suggest a watchful watchmaker.

After Cindy Bennett's surgery, she immediately wanted to adopt. Her therapist said she needed to grieve the loss of her fertility first. After a year of painful soul-searching, she felt she was ready.

The process of adoption can take years, but Bennett's brother, a lawyer in Florida, knew of a pregnant woman in his state who wanted to give up her baby. The Bennetts flew down to meet the birthmother, Cindy cried with her, and the woman selected the Bennetts to raise her son. A month later she gave birth. As soon as Cindy and Randy came home with Jackson, 'it didn't matter anymore that I didn't actually give birth to him', Cindy says. 'He just felt like ours.'

'He healed my soul', she said. 'I mean it sounds so hokey. But when you're a woman who can't give birth and you start to doubt yourself on so many levels, then you get this child, everything just changes. It just heals you.' Among Cindy's friends, Jackson became the poster child for adoption; the match seemed perfect. Eight years later, the Bennetts adopted another son, Carson. The boys, when I met them, were seventeen and nine. Jackson is self-assured and shares Randy's sarcastic wit. Carson is carefree and all smiles. 'I honestly just couldn't imagine being a parent of any other children than my children', Cindy says. 'They were destined.'

Miriam Klevan heard similar remarks from the infertile couples she interviewed. They had all adopted, and in her analysis of the transcripts, 'it immediately stuck out that people were talking about fate', she says. More than half the thirty-eight parents expressed a sense of destiny. Klevan has two adopted children of her own, and she says for the first year of her son's life she and her husband would lie in bed every night and say it's like he's meant to be theirs. 'And when I started doing these interviews, I came home and I laughed and I laughed, and I said, "You and I thought we were so special."' The adoption professionals

I spoke with also told me that adoptive parents feel their children were destined to be theirs more often than not.

Some adoptive parents will forgive whatever brought them to where they are. One couple who talked to Klevan had a baby boy at home with them for ten days before the birthmother changed her mind and took the child back. 'They were devastated', Klevan says. 'They closed the door to that child's room and didn't open it for six months.' But the father said it was supposed to happen so that they could adopt the son they adopted later. A number of adoptive parents say their painful infertility was meant to be. 'Once we had the babies in our arms, we got to the other side', Cindy told me. 'To me, it made sense. It's like, *Oh okay, now I get it. That's why I had to do this, this, and this.*' One mother, a self-proclaimed atheist, quoted me a country song: 'God bless the broken road that led me straight to you.'

||||||||||||||

Why do we so quickly jump to explaining a series of events as necessary steps leading to a preordained destination? It all begins with our mind-reading abilities. As we saw in the last chapter, within a few months of birth, babies start processing the world in terms of goals and intentions. We easily see purpose behind not only the creation of objects but also the induction of events. Most of the time we dismiss such attributions, but unexpected and personally significant events are not so readily marked up to happenstance. They seem purposeful. By definition, significant events are occurrences that serve some important function in our lives, for better or for worse, and when something serves a function, we're prone to wonder if the function was imparted intentionally.

A surprising and significant event also demands deep reasoning about its causes because we want to get to the bottom of what's happened – we want to understand how to prevent it from happening again (if it's bad) or reproduce it (if it's good). In a study of people who had lost a spouse or child in a car crash, 85 percent of surviving spouses and 91 percent of parents asked themselves, 'Why me?' or 'Why my [spouse/child]?' They understood that people die in crashes, but they wanted to know why their families were specifically targeted. An explanation of 'these things happen' provides no satisfaction and no sense of control. But when a mind is behind an event – whether it's the mind of a person or God – you just might be able to influence or at least predict the mind's future behaviour. Therefore, we're better off with a bias that assumes and psychoanalyses intentionality everywhere, just in case. When it comes to physics, we are all sometime psychologists.

There may be another reason crediting an event to a mind, rather than to chance or physics, provides explanatory satisfaction – why it scratches that deep *why* itch. A popular anecdote has a scientist lecturing on the universe when an old woman calls *hogwash*. 'The world is flat and rests on the back of a giant turtle', she says. 'But on what does the turtle stand?' the scientist replies with a smirk. 'I see where you're going with this', she says. 'It's turtles all the way down.' Many atheists argue that belief in God doesn't explain the origin of the universe or events therein, because what explains God? Is it gods all the way down? If anything can come from nothing, why not assume it's the universe and cut out the middleman? Very logical! But not very satisfying. Divine will is not just any old explanation that needs another turtle under it. 'God is a magical cause, and so you don't need to explain further', Jesse Preston, a psychologist at the University of Illinois at Urbana-Champaign, told me. I asked her why agents (such as God) don't need causes. 'Part of it, I think, is our understanding of action', she said. 'We see the action of agents as being caused by itself. We have this idea of free will, and I can't find anyone who believes in free will who can explain that further. It's free, right? That's the whole meaning of it. There it is.' So we have this very familiar and acceptable notion of uncaused causes, in the form of free will. If we can attribute an event to an agent's free will, it shuts down that infinite regress of explanations and it scratches that itch – even without really explaining anything. Free will is the bottom turtle, and that's that.

But here I've offered two seemingly incompatible accounts of why resting an explanation on an agent satisfies us. To the degree that we lean on the presumed predictability of an agent that we can psycho-analyse or placate, we can't also lean on the inherent *un*predictability of its freedom of will. And yet my hunch is that we do lean on both. How is that possible? I have no solution except to say that, perhaps in our everyday thoughts about causes, we don't think deeply enough to notice the logical contradiction. Satisfaction isn't always about logic, after all.

ıııııııııııııı

Even atheists, who tend to pride themselves on their powers of reason, often abandon logic when trying to make sense of their lives. Bethany Heywood, in collaboration with Jesse Bering at Queen's University Belfast, used online instant messaging to interview thirty-four people who said they didn't believe in souls, spiritual forces, or gods. She asked them to recall a learning experience and a low point in their lives, then asked follow-up questions about those experiences. Heywood found that the majority gave at least one teleological response or one con-flicted response – an answer indicating previous or hesitant teleological reasoning. For example, when asked if he deserved to fail a course, one typed, 'I don't know, maybe it happened for a reason . . . so that I could see that even if I failed a course, my life wouldn't actually end.' Another, when asked if getting a dream job was 'supposed to happen', typed, 'Yes. Even though I don't really believe in fate and such woolly nonsense.' (This response was coded as 'conflicted'.)

When asked if thanking fate for one's adoptive children is a mat-ter of religious faith, Adam Pertman, the executive director of the Evan B. Donaldson Adoption Institute, said, 'I think this is irrespec-tive of whether you're devout or an atheist or agnostic. I think the sensibility arises from the magnetic attraction to your child.

'And thank God it does', he went on to say, 'because then we gotta put up with them as teenagers.'

The Hot Hand of God

Some things show evidence of design through the function they serve. Others, through their display of order – order that is often illusory.

In June 1944, Germany began launching its V-1 Buzz Bombs at England from the coast of France. Over the coming months, thousands of these bombs with wings and jet engines came raining down on England, killing about 6000 people in London. Maps of the city showed clusters of impact points, leading people to wonder about the patterns. Were certain parts of town targeted more than others?

Finally, a statistician took a careful look at the damage. He sliced the city into squares and counted how many Doodlebugs (as they were also known) had landed in each square, then compared the distribution to what one would expect if they fell randomly. He found almost no difference. People had given the Germans much more credit for their targeting ability than they deserved.

The English had fallen prey to the *clustering illusion* – the tendency to find significance in incidental clusters of random data points. They had also demonstrated the *Texas sharpshooter fallacy*, named for the apocryphal Texan who shot up the side of a barn and drew bull's-eyes around the clusters of bullet holes: they'd drawn bull's-eyes around the bomb clusters and inferred terrific aim. A real-life German smart-bomb fallacy.

Apophenia is the general term for seeing illusory patterns in information. (Michael Shermer, the founder of the Skeptics Society, coined the more self-explanatory synonym *patternicity*.) We notice faces in the clouds (see chapter 6), we connect lucky charms to successful outcomes (see chapter 3) or wishes to wishes-come-true (see chapter 4), we see conspiracies among our competitors, and we hear messages in records played backwards. (At university my roommate discovered that if you record yourself singing, 'Ooh you sniff turkey fat' and play it backwards, it sounds just like, 'Happy birthday to you'. Or maybe he discovered it the other way around . . .) Apple ran up against apophenia when it introduced the shuffle feature on its iPods. Listeners complained that it would play strings of songs by the same artist, that it wasn't really random. (Some also saw something spooky in its occasional matching of a perfect track to the situation or to the listener's mood.)

People are notoriously bad at recognizing and producing randomness. Psychologists have repeatedly demonstrated what's called the *gambler's fallacy:* after a string of identical results, we expect a different result to follow. Flipping a fair coin, two or three or four heads in a row happens frequently, but to the untrained eye they look suspicious. Apple, in response to customer complaints about the apparent lack of randomness on their music players, introduced a feature called 'smart shuffle' that allowed users to manually avoid repetition in their playlists. Steve Jobs announced, 'We're making it *less* random to make it feel *more* random.'

Our mistaken belief in the so-called hot hand in basketball, where a player makes a series of scoring shots, may result from our seeing streaks in mutually independent shots and assuming the runs couldn't have happened by chance. The player must have some kind of momentum. (Shooters actually have a *decreased* chance of making a shot after three or four successful baskets, either because they're taking riskier shots or because they've drawn increased defensive pressure.)

The bias that leads to apophenia, the spotting of illusory patterns, is a feature, not a bug. There is a fair amount of order in the world, and it would behoove us not to miss it. The tides come in, the tides go out; eating an apple usually doesn't kill you; and objects fall down rather than up when dropped. We learn and survive by finding patterns in the environment. Without the ability to note regularity in spatial arrangement or cause and effect, the world would dissolve into a terrifying sea of chaos. And, as with bears and boulders, we're typically better off spotting patterns that aren't there than missing ones that are.

In addition to our bias for pattern finding, we also have one for believing patterns were created by agents. George Newman and colleagues conclude in one paper that 'by at least 12 months of age, infants appreciate that agents are capable of creating order, while inanimate objects are not.' These two biases together lead us to see the world as uncannily, and intentionally, ordered. Someone somewhere obviously has both a game plan and a hot hand.

A Magical Kingdom

A week before their wedding in 2002, Alex and Donna Voutsinas were paging through family photo albums when they came upon a Polaroid snapshot of Donna, age five, posing with a character at Disney World. A man in the background with dark hair caught Alex's eye. Not the one wearing short shorts (this was 1980 after all), but the one with the stroller. It was Alex's father. And there, in the stroller, was Alex.

The two children, each visiting the Magic Kingdom for the first time – Donna from Florida and Alex from Canada – would not meet for another fifteen years. When they saw the photo, 'I got chills', Alex told a reporter. 'It was just too much of a coincidence. It was fate.'

Coincidences, as I noted in chapter 4, are the manna of magical thinking. They rain down on us daily and feed our ravenous appetite for meaning. But they're not so special. Because of the sheer number of things that happen in life, some of them are bound to be extraordinary. Millions of people visit Disney World each year, and each person takes I-don't-know-how-many photos. The fact that a future spouse was caught in friendly fire does not surprise me. (Even less spooky a coincidence is the appearance of 'It's a Small World' references in nearly every newspaper account of the Voutsinas photo.) Yet, coincidences grab our attention. And we crop, magnify, and enhance them in the stories we tell others and ourselves.

We often see coincidences as order orchestrated from above. Many suggest a designer because they present an improbable pattern in the world. Others link the outside world to our thoughts – an event holds special meaning for us. *Referential thinking*, the finding of self-relevant meaning in random events, is a natural part of our stream of consciousness, according to Laura King of the University of Missouri at Columbia. 'We've all had that experience where it's the one day you really needed to get up, and your alarm clock didn't work, and it's like, "*of* course"', she says.

The Swiss psychoanalyst Carl Jung used the term *synchronicity* to describe meaningful coincidences. Many synchronicity enthusiasts keep coincidence journals, noting the circumstances and meanings of odd occurrences in their lives. They hope that by tracking such phenomena, they can interpret them and also learn to better detect them and bring more synchronicity into their lives. Perhaps by decoding the universe's subtle (or not-so-subtle) messages, one can find spiritual guidance. I see synchronicity's usefulness in its manifestation not of a higher consciousness but of one's own subconscious. Each remarkable moment is like a Rorschach test: how will you interpret it, and what does that say about you?

A man sees his shrink, who shows him inkblots. To the first, the man says, 'It's a couple having sex, missionary.' To the second, 'Sex, doggy-style.' To the third, 'Reverse cowgirl.' (Tailor the telling of this joke to your preferred level of filth, from what we have here up to *The Aristocrats*.) 'It seems you've an obsession with sex,' the shrink says. '*Me?*' the man replies, '*You're* the one showing me the dirty pictures.' As with reading fate into synchronicity, the man is projecting his private interpretations onto the designs of an external agent. But there's no kink in the ink; it's in him.

People tend to seek advice when they're standing at crossroads in their lives, but reading coincidences as signposts can be dangerous. Mark David Chapman doubted himself every step of the way before shooting John Lennon, but he kept seeing signs that urged him on. In New York City, Lennon lived in an apartment building called the Dakota. Exterior shots of the building appeared in *Rosemary's Baby*, the 1968 horror film in which a woman gives birth to a devil child.

Rosemary's Baby was directed by Roman Polanski, whose pregnant wife, Sharon Tate, was murdered in 1969 by followers of Charles Manson, whose mayhem was inspired in part by the music of the Beatles. (Manson's followers would sometimes write 'Helter Skelter', after the Beatles song, on walls using their victims' blood.) As Chapman stood outside the Dakota apartments on 8 December 1980, pondering these associations, the actress Mia Farrow walked by. Farrow played Rosemary in Polanski's film. Chapman decided he had to kill Lennon that day.

In select cases, attributing a coincidence to a cosmic consciousness can be justified. Not scientifically, but pragmatically; it's the healthiest option. One of Klevan's couples learned through their agency that two little girls had become available at an orphanage in India. They couldn't decide between the two and asked to adopt both, but were not allowed. The decision tortured the wife – it was a Sophie's choice: one child would be lost to her – and she finally selected the younger child. Someone from the agency brought both girls over on the same plane, and waiting for the older girl at the airport was an Indian couple who'd lost a two-and-a-half-year-old daughter to leukaemia. The Indian couple's new daughter? Also two and a half. It wasn't a remarkable coincidence, but to the woman who'd selected the younger girl it meant she'd made the right choice. The older girl was supposed to go to the other couple, and it alleviated her guilt. Klevan says adoptive parents use coincidences as signs 'all the time'. What if the woman had selected the two-and-a-half-year-old? Would she have been racked with regret? Perhaps. Or she might have found some other coincidence – a similarity in names or birthdays – to justify her decision. And who could fault her?

Relying on coincidences to plan future efforts (rather than to justify past actions, as this woman did) can work out too. While relying on tea leaves will often lead you astray, at forks in the road where you can't go wrong, coincidences can provide the conviction you need to overcome your indecision and pick a path, any path. But Donna Voutsinas, from the Disney photo, realizes the inherent dangers of relying too much on coincidence. Speaking of her husband, she told a reporter,

'I was glad he proposed before [we found] the picture, because I know that it's because he loves me and not because he thought it was meant to be.'

The Alchemy of Creativity

With his compulsive pattern finding, Mark David Chapman showed evidence of schizophrenia. A nurse who wrote about her own experience with paranoid schizophrenia illustrates pattern finding at its extreme. 'Every single thing "means" something', she recounts. 'I have a sense that everything is more vivid and important; the incoming stimuli are almost more than I can bear. There is a connection to everything that happens – no coincidences.'

If there's one chemical most responsible for magical thinking, it's the neurotransmitter dopamine. Levodopa, a drug used to treat Parkinson's disease, is converted to dopamine in the body and can lead to psychosis. And drugs that block dopamine receptors reduce delusions and racing thoughts in schizophrenia patients.

Peter Brugger, the head of neuropsychology at University Hospital Zurich, has spent several years researching the neurological basis of magical thinking. In one study, he and collaborators briefly showed images and strings of letters to believers in the paranormal and to sceptics. Believers were more likely than sceptics to judge letter strings that were not words as words and to see jumbled facial features as regular faces. In another study, subjects saw two unrelated words and had to think of a third word that could connect them. Believers responded faster than sceptics. In general, magical thinkers appear to see meaningful patterns and associations where others don't. But giving levodopa to sceptics makes them behave more like believers.

The nurse's description continues, 'I feel tremendously creative.' This sentiment – along with the lives of such luminaries as John Nash and Vincent van Gogh – hints at the similarities between schizophrenia, magical thinking, and creativity. Those who see connections that others miss sometimes hit upon valuable new ideas. A surprising punch line, a useful invention, an illuminating metaphor. The same pattern finding and association making that lead to belief in the paranormal also allow for innovation in physics, painting, and poetry. Magical thinking can aid in the process of creation. But then you've still got to sort the wit from the chaff, the good ideas from the bad – which requires turning down the magical thinking to critically analyse what you've got. Write high, edit low.

Magical thinking is also important for letting loose and having a good time. Brugger finds a positive correlation between magical ideation and the ability to find pleasure in life. More magic, more fun. (As long as reality stays within arm's reach.) 'Those students who are not magical are not typically those who enjoy going to parties', he says. 'To be totally unmagic is very unhealthy.'

Love the One You're With

Patterns suggest a pattern maker in part because of their improbability. Events that seem too good to be true also seem improbable and thus intentional.

Almost no adoptive couple ever turns down a child allocated to them, Signe Howell, a social anthropologist at the University of Oslo who studies international adoption, wrote in her book *The Kinning of Foreigners*. Instead, they feel an instant bond: 'Again and again, adoptive parents have told me how they experienced an immediate sense of fate upon being told about a specific boy or girl who has been allocated them. They are overcome by a profound sense of destiny that has connected them and the unknown child.'

This love at first sight resembles the results of a set of experiments conducted in the 1940s at the University of Pennsylvania. Two objects are placed in front of a child. If the child is asked to say which one he would prefer as a gift, each object has an even chance of being picked. But if he records his preference *after* randomly receiving one of the items, the odds are great that he'll prefer the gift he received. This finding has been repeated numerous times over the years and has been dubbed the *mere ownership effect*. Because of our desire to keep up a positive self-image, we automatically overvalue anything associated with ourselves, including things we merely own. I believe the endowment of a child, or of any life circumstance, also enhances our evaluation of him-her-it. We judge our own reality in the best possible light.

Another basic psychological process at play in the appreciation of one's fate is the *existence bias*. We prefer existing states of affairs over alternate realities just because of the mere fact of their being. Whereas the mere ownership effect is motivational (we want to feel good), the existence bias is purely cognitive; we tend to judge extant entities to be 'good' and 'right' and 'the way things ought to be' whether or not we have any stake in their existence. Of course, the bias has its limitations – we don't cherish every ugly feature of the world (cancer, the 1998 BMW M coupe) – but it gently steers us away from starting a revolution every time we feel antsy about the status quo. Things can't always be that bad.

For the most part, these (and other) optimizing processes are invisible to us. Someone hands you one of two similar gifts, and instead of thinking, *I guess I'll go ahead and prefer this one since I now own it*, you think, *how did she know to pick that one?* (On my second birthday, my reaction to each gift was, 'Just what I always wanted if I ever knew about it!' My family still teases me.) Daniel Gilbert and collaborators found that when people judge a selection of options after (versus before) being assigned a particular one, not only do they rate their assignment more favourably (that's the mere ownership effect), but they also give kudos to the person who made the selection. What a knowledgeable, capable, and kind gift giver! Gilbert and team wrote, 'Technically speaking, participants confused their own optimization of subjective reality with an external agent's optimization of objective reality. Simply speaking, participants mistook "the magic in here" for "the magic out there".'

So we make lemonade out of lemons but without realizing it, and as a result we think life is miraculously handing us the lemonade, sweetened to taste. What are the chances that the universe would cater so well to us – that, say, the *perfect* baby would land in your lap? Pretty slim, unless someone's pulling some strings.

Parallel Universes

In discussions of life's unpredictability, the humble butterfly often plays an outsize role. In the 1960s the mathematician and meteorologist Edward Lorenz was running computerized weather simulations when he noticed that tiny differences in the initial state of a simulation led to dramatically different outcomes. A small change became amplified over time. His colleagues soon concocted the image that has stuck with us: a butterfly flapping its wings in Brazil can set off a tornado in Texas.

Life is an agglomeration of butterfly effects, contingency balancing on contingency. Your very existence depends on a single sperm cell veering right instead of left for a split second. And that sperm's existence depends upon your parents happening upon each other, perhaps nearly missing. Maybe they met on the underground after the previous train escaped one of them by a hair. To think of all the ways you could be somewhere else, or someone else, in an unrecognizable universe – to grasp that human history owes its particular path to the wings of a fly on a Cro-Magnon's ass – could make our current reality seem meaningless, just one of innumerable possibilities, a winner by blind chance. But does it?

Imagining alternate realities – picturing what might have been – is called *counterfactual thinking*, and it's a ubiquitous element of cognition. Children do it by the time they're two, as soon as they can say 'if only'. Playing out different scenarios in our heads helps us piece together causes and effects, and it allows us to learn from our mistakes so in the future we can narrow the gap between what is and what should be. 'As long as we are human beings who strive toward goals, we'll have counterfactual thoughts to tell us how we could have achieved those goals', says Neal Roese, a psychologist at the University of Illinois at Urbana-Champaign. We also sometimes think about how things could have gone *worse*, in order to console ourselves after a loss or congratulate ourselves after a win.

Roese and several collaborators, led by Laura Kray of the University of California, Berkeley, wanted to know the effect of might-have-been musings on attributions to fate. They asked a group of subjects to identify turning points in their lives, then asked a subset to imagine life minus the defining incident. But considering how things could have turned out differently didn't highlight the haphazardness of their lives; instead, it made them believe more strongly that the episode was a product of fate. They appeared to be reasoning that their current lives were actually somewhat improbable and so couldn't have happened by chance alone.

Because counterfactual thinking is ubiquitous, the nagging feeling that what's happening is improbable and thus artificially selected may be unavoidable.

Thank Your Lucky Stars

One morning in August 1944, a German Doodlebug exploded in London, disturbing a butterfly and causing it to flap its wings. No one seemed to notice the tiny breeze.

A year later, on the morning of 9 August 1945, the wings of *Bockscar* lifted it into the air. The B-29 bomber loaded with a five-tonne atomic bomb named 'Fat Man', took off from Tinian, an island 1500 miles southeast of Japan. The US had dropped 'Little Boy' over Hiroshima on

6 August, immediately killing tens of thousands of people, but Japan had not yet surrendered World War II. Around 9:30 AM, the weather scout plane *Up an' Atom* reported a few clouds but decent conditions over the next target. Clear for bombing.

Oh, but what's that? The year-old turbulence of a butterfly half-way around the world? By the time *Bockscar* passed over its target at 10:44 AM, the city was covered in haze. According to the pilot's flight log, 'Two additional runs were made, hoping that the target might be picked up after closer observation. However, at no time was the aiming point seen.' So the crew left the city of Kokura and made their way over to their second choice, Nagasaki.

Most people have never heard of Kokura. In this regard, it can be counted as one of the luckiest cities in the world.

||||||||||||||

We're fascinated by near misses. In 1993, a German motorcyclist hit a truck, went flying into a tree, and was impaled on a branch. The head-line in a Norwegian newspaper, next to a photograph of this unfortunate gentleman with a branch still going into his chest and out his back, read, 'Verdens Heldigste', or 'World's Luckiest'. Now, before you protest that perhaps I have made a spectacular error in translation, let me add the ending: the man survived. The branch had missed all his vital organs.

Luck is as tricky to define as it is to tame. As a first pass, you might calculate an outcome's level of luck by multiplying its positivity by its improbability. Rare good things are lucky. But what about Kokura? Nothing positive happened to it. In fact, it lost thousands of neighbouring countrymen. And the German motorcyclist had surely seen better days. 'Luck' is often yoked to terrible, terrible things.

Karl Teigen, a psychologist at the University of Tromsø in Nor-way, has spent years studying what we mean when we talk about luck. One conclusion he's reached is that, on par, 'lucky' events are not pleasant. In a search of newspaper stories, for example, he found that 'with the exception of an occasional sports champion and a double lottery winner, the typical lucky person had survived a plane or car crash, had been stabbed or shot, fallen off a cliff or bridge, been shipwrecked, or surrounded by flames'. In a one-month period, he found one mention of 'bad luck', in a story about a soldier who had stepped on a mine – but the soldier had nevertheless commented on his own good fortune in losing just the one leg.

Teigen concluded that luck derives not from the absolute value of an outcome but from its relative value – we assess luck by standing reality next to its most salient counterfactual neighbour. If you lose a leg but think of how you almost lost two, you'll consider yourself lucky. If you win second prize in a lottery, but miss first prize only by changing a digit at the last minute, you'll consider yourself unlucky. Why wouldn't you compare losing a leg to not stepping on the mine at all? Mentally altering an effect or a recent cause in a chain of events comes more naturally than mentally undoing an earlier cause. Once you've skewered yourself on a tree, you'll tend to think of the untouched organs right next to the branch hole, not about how this never would have happened if you hadn't taken the extra five minutes to floss this morning.

Luck isn't a direct product of improbability, either. In one study, Teigan divided a roulette wheel into three wedges of equal size: one red, one yellow, and one blue. He divided another into eighteen sections of equal size: six red, six yellow, and six blue, interspersed. Landing on red wins. Eighty-five percent of subjects felt it was luckier to land on red on the wheel with eighteen divisions than the one with six, even though the probability was the same on both wheels. On this wheel, whenever you land on red, the slice is so narrow you're almost landing on blue or yellow. It's a close call every time. Similarly, of two boys standing near a large icicle that falls and lands between them, people judge the closer boy to be luckier, not despite the fact that he was closer to death but because of it.

Teigen concluded that the two factors defining luck aren't positivity and improbability but rather the difference in value between reality and its most salient counterfactual, and the closeness of that counterfactual. How much better or worse things could have been, and how nearly they were. The chance of a cloud layer over Kokura wasn't remarkably slim, and the city looked pretty much the same on 10 August as on 9 August 1945. Kokura's luck comes from how easily we can imagine the clouds not being there and from how big a difference those clouds made.

And luck is not an impersonal impetus. Teigen has found that feeling lucky correlates with feelings of gratitude, a distinctly social sentiment. And the thanks are often of an existential kind, directed not towards a person but towards God or the universe or fate.

A few years ago, Eugene Subbotsky, a psychologist at Lancaster University who studies magical thinking, was strolling through Moscow with his young son, with no one around. They walked past an empty, parked car. 'Just when we were passing by, the engine started', Subbotsky told me soon after the event. 'My son said "beebee" – he calls cars beebees, he's only three years old – "Papa, beebee." And then I look at his eyes, and I see his eyes widening.' The car starts moving. It swerves towards them. Finally, it turns a little more and hits an iron gate a few inches away. 'We escaped death very narrowly', he says. 'I could have been smashed to pieces with my little son. I am a rational man, I am a scientist, I'm studying this phenomenon, but there are some events in your life that you cannot explain rationally. Of course you can always write it down to chance and say, "Okay, it's a coincidence", but it's such a rare coincidence that you start thinking mystically and magically about things.'

Subbotsky felt lucky and he felt grateful. 'Under certain circumstances I really feel like someone or something is guiding my life and helping me', he says. Well, that's one interpretation. Personally, I might have supposed that whoever was guiding my life was trying to kill me and needed to work on his aim.

But I'm not going to stop Subbotsky, or anyone else, from feeling thankful. Taking time to give thanks each day has been shown to increase happiness, optimism, sleep quality, and generosity towards others. Whether you're prone to say 'Thank God' or 'Thank goodness' in response to good fortune, gratitude is a natural response, and it's an emotion worth cultivating.

On a Mission from God

Telling stories of fate requires connecting two or more events together. *This happened so that that could happen.* The second event makes the first event seem purposeful.

People can also feel a more general purpose to their entire lives by stringing many episodes together.

We define ourselves using narratives, according to Dan McAdams, a psychologist at Northwestern University and a leading researcher in 'narrative psychology'. 'To be an adult means, among other things, that you've figured out how you got to where you are and where you're going', he told me. 'And the only way I know of to convey that to another person is through a story.' The kind of story I told about myself in the introduction of this book, for instance. People also explain their pasts and futures to *themselves* in terms of narratives. 'These stories function to give life a certain level of coherence', McAdams says. Lives comprise many unrelated events, but our pattern-finding minds draw connections. We revise our histories to make better stories, omitting certain details and changing others so they conform to a cohesive, compelling, and often complimentary tale.

The series of events in your story comes to define your identity, and you wouldn't want to change any one of them because to be someone else would feel inauthentic. How lucky you are that university A did not take you, because you owe your career, your best friend, and many life lessons to your attendance at university B. Would you trade all those in?

We compose our life stories using the data given – the somewhat random happenings of our pasts – but then we get the roles of the data and the interpretation confused: we stare in wonder at how well the events seem to fit the theme, forgetting that we custom-fit the theme to the events. It's another example of the Texas sharpshooter fallacy, but instead of drawing a target around a cluster of bullet holes and gawking at the aim of a marksman, you're constructing a story around a series of occurrences and marvelling at the insight and wisdom of providence. One stray bullet and you wouldn't be who you are today.

For many people, a dramatic change in one's life, perhaps taking on a new role – such as parenthood – provides a moral or a theme to one's story that suddenly contextualizes everything that came before. One parent wrote on an adoption board, 'I've been filled with amazement at how my life seems to be falling together since hearing about [the girl we ended up adopting]. For the first time in my life, everything seems to make sense. It's like there were a few hundred curvy rows of dominoes, and they all seem to be merging into a single line.' In retrospect, all those dominoes were leading up to something.

One can certainly tell a good story of one's life without resorting to unnatural intervention. Event B just logically followed event A. You decided X because you wanted Y. But it's those crucial turning points where things could so easily have gone a different direction – the ones that invoke counterfactual thoughts – that arouse a sense of guidance. Some people would prefer to attribute these defining incidents to rolls of the cosmic dice, but perceiving fate can inject one's narrative with greater meaning. Laura Kray and colleagues found in their research on counterfactual thinking and fate perception that those who considered alternative outcomes to their lives not only saw the real outcomes as fated but also more strongly agreed that 'It made me who I am today' and 'It gave meaning to my life'. Further, the data revealed that the sense of meaning was brought about by the perception of fate. 'When people believe their lives are as they were meant to be', the authors wrote, 'they experience the gratification of being on the "right" course and fulfilling their life's mission'.

Laura King has found that, when combined with positive affect, referential thinking (seeing events such as traffic light changes as having a significance meant specifically for oneself) is correlated with finding one's existence purposeful and meaningful. 'It makes events that are random and chaotic feel like they belong in a coherent narrative about who you are and what your life is', she says. 'You become a character in a much more interesting narrative when the whole world is conspiring to take part in your personal drama.'

You want to believe that all those flukes of luck leading to where you are were somehow meant for you. Customized kismet means someone's got your back. It also means that those events that happened for a reason may be building up to some future purpose. It gives the entire story of your life both continuity and a destination, something to strive for. You were put here for a reason, you matter, and you're on a mission. Everything before now was to prepare you for your calling! The universe is counting on you! Now hop to it!

Sometimes it's fun to pretend.

The Holy Scapeghost

The existence of suffering in the world has made people question their faith for millennia. And yet many believers, rather than reject their faith, attempt to understand the purpose of the suffering. They see it as part of God's plan.

A disastrous event may even enhance one's belief in a master planner. In the last chapter, I mentioned the dyadic template of morality: for each recipient of help or harm, there is someone doing the helping or harming. Kurt Gray and Daniel Wegner argued that we make liberal use of the template: if there's no obvious responsible party, we find a scapegoat. But what happens if no acceptable scapegoats are in sight? We credit a supernatural one.

Gray and Wegner presented subjects who believed in a higher power with one of four stories. In all four versions, a family is picnicking in a valley when the water level rises. In half the stories, lunch is ruined by the flood, and in the other half lunch is *really* ruined because everyone drowns. Also, in half the stories a dam worker is said to have caused the flood, and in half of them the cause of the flood is unknown. Subjects then rated how much the story's outcome was part of God's plan. God drew much more blame when people died and no one was clearly responsible than in the other three scenarios. The tragedy needed an explanation, and human intervention wasn't an option.

We've seen similar stories in the news. Both Christian and Muslim leaders claimed the 2004 Indian Ocean tsunami that killed a quarter of a million people in Indonesia was punishment from God. After Hurricane Katrina flooded New Orleans in 2005, the city's mayor said God was 'mad at America' for the Iraq war, while a Christian group noted a resemblance between a satellite image of the storm and an ultrasound image of a foetus, suggesting that Katrina came to avenge the 'ten child-murder-by-abortion' centres in Louisiana. The conservative commentator Glenn Beck called the earthquake and tsunami that hit Japan in 2011 a 'message' to humanity. And of course Noah would have something to say on the matter of floods. It's no coincidence the legal term for an unpreventable natural disaster is an 'act of God'.

We might even give more credit to a higher power for bad stuff than for good stuff. Carey Morewedge has found a *negative agency bias* – a tendency to attribute negative events more than positive events to agents. In one of his studies, subjects played a game of chance where an impartial other player could secretly intervene on some rounds of the game. Subjects were most likely to suspect the other player had stepped in when they lost money, particularly when they lost big.

If you're one to see the good in people, and assume others are too, you might look askance at this negative agency bias waltzing in here and telling you that looking askance is actually something you do quite often. But wait – it can explain! Recall how our favourite types of explanations are those involving intentional agents, rather than chance or impersonal causality? Well, we tend to think more about what causes negative experiences than positive experiences, in part because they're more unexpected and in part because they mean something's wrong and needs fixing. And more explanation means more attribution to an agency. So it makes logical sense that good fortune goes unquestioned but disruptions become whodunits. In research by Ronnie Janoff-Bulman and collaborators, everyone in a sample of paralysis victims asked 'Why me?' and all but one settled on an answer, but only half of a sample of lottery winners asked and answered the question. (Among the twenty-nine victims, ten blamed God, seven credited predetermination – a more implicit version of a divine plan – six cited benefits to the ordeal – invoking teleology – and two noted moral deservedness – and thus karma.)

The negative agency bias also results from the fact that we like to take credit for positive outcomes and blame failures on someone else.

Gray and Wegner suspected that if acts of consequence, particularly unfavourable consequence, arouse feelings of divine authorship, then those who suffer the most should have the most reason to believe in God. They propose that people believe in a 'God of the moral gaps' who fills in when mortal scapegoats are unavailable. So they plotted fifty points on a chart, each representing one US state. One dimension represented 'misery index' scores accounting for such factors as rates of infectious disease and infant mortality, and the other dimension represented the strength of citizens' religious beliefs. The pattern was clear; they found a strong positive correlation between misery and the Almighty, even controlling for income and education. The more 'acts of God', the more God. (If you ask people explicitly whether a bad event or a good event is more likely to be caused by God, those who believe in a loving God will probably upon reflection say the good event, but the bad event is still more likely to arouse spontaneous blame.)

Gray and Wegner offer an alternative account for their misery-index findings: suffering leads people to seek comfort in a higher power. They suggest the needs for comfort and blame likely work in tandem: 'God may be both the cause and cure of hardship', they write. Homer Simpson demonstrated his understanding of this funky logic when he gave his most famous toast: 'To alcohol! The cause of, and solution to, all of life's problems.'

Good Grief!

With God's rap sheet full of floods, infertility, and that life-ruining prom-night pimple, one could be forgiven for wondering why we still see him as a nice guy – why, when Hollywood needs a fill-in, they look to Morgan Freeman over Al Pacino. (See: *Bruce Almighty*; *Devil's Advocate*.)

Most people are pretty resilient. Studies find that the majority of trauma survivors find good things that came out of accidents, bereavement, cancer, combat, disasters, genocide, heart attacks, HIV infection, house fires, and incest (just to pick on the first third of the alphabet). Sometimes the benefits are measurable, in insurance money or spare time or career paths or friendships you wouldn't have formed otherwise. Benefits can also take the form of an improved self. After examining the silver-lining literature and surveying people who'd recently encountered suffering, Richard Tedeschi and Lawrence Calhoun enumerated five distinct forms of post-traumatic growth: a deepened appreciation for life, enhanced ability to relate to others, a sense of new possibilities, intensified spirituality, and greater personal strength (*whatever doesn't kill you . . .*).

Sometimes the benefits even outweigh the losses; the silver lining somehow shines brighter than the sun that was obscured in the first place. Cindy Bennett describes her cancer as a net plus. It brought her Jackson and Carson, it led to her current business (she started an infertility site, sold it, and founded an online marketing company – named after the willow tree planted for her when she had cancer), it increased her empathy and appreciation for life, and in the end it even strengthened her marriage. Her experience with cancer and infertility was 'a gift to make me who I am', she said.

In the research with paralysis victims by Janoff-Bulman, subjects made comments such as, 'I was moving too fast at the time, and I think this was the best way to slow me down', 'I was leading a sheltered life, I suppose, compared to what it is now. Now I'm just in a situation which I enjoy', and, 'I see the accident as the best thing that could have happened 'cause I was forced to decide my faith'. And when asked by Miriam Klevan to mentally rewind their infertility and adoption and to picture having as many biological children as they'd like, parents always said they'd be worse off. 'One refused to answer and said, "I can't even imagine that"', Klevan says. Our motivation and ability to find benefits in our present circumstances, combined with teleological reasoning, may explain why most people think so positively of the higher powers operating in their lives: if everything was meant to happen, and if in the long run most things are good, then God must be great.

That doesn't mean he can't also be infuriating. Pointing fingers after a tragedy makes sense, according to Omar Sultan Haque, a researcher at Harvard. The immediate need after any offence is to hold someone responsible and prevent further injury. 'You wouldn't expect someone upon first seeing an accident to say, "I wonder what the larger purpose or meaning of the accident is"', he says. 'They're going to say, "Well, who was driving?"' Only after ample reconsideration and careful soul-searching do we, sometimes, retract the middle finger and graciously give a thumbs-up. The accident was for the best.

So post-traumatic growth can add a rosy tint to our God goggles, and the influence is bidirectional: seeing a benevolent deity behind hardship can aid post-traumatic recovery. Even if you don't know what God's up to, you trust he's not up to no good. Let's say you receive a questionable present from someone who's otherwise known for her insightful gift-giving. You'll probably give the gift a chance and attempt to find something to like about it. Perhaps the plushy chandelier will finally bring out the fuchsia in your drapes. (Your friend's got an eye, you know.) Kenneth Pargament of Bowling Green State University in Ohio has found that interpreting a recent negative event as part of a loving god's plan enables healthy coping. The subjects in his research sought and managed to extract benefits from their adverse experiences; they expected them to be in there somewhere.

People of faith (or even suspicion) realize that God works in mysterious ways, but that after a little legwork they're sure to solve the mystery and find the hidden treasure. The booty, rife with life lessons, new appreciations, and other valuable goodies, will in fact be self-manufactured, but to call it fool's gold would be missing the point.

Karma Suits Ya

One of the subjects in the paralysis study said he injured his spinal cord in a car accident because he deserved it. After acquiring Cadillacs and jewellery, he said, he should have moved out of the ghetto so as not to arouse jealousy. 'And then for number two', he went on, 'I shouldn't have had [redacted] in my car. Then, I had a good woman at home, and I was dogging her, you know, pimping was involved . . . you just can't have your cake and eat it too. And if you do wrong, you reap what you sow.'

One way we make sense of accidents is by finding a reason the accident deserved to happen. Piaget called such karma *immanent justice* and noted that young children reason in terms of 'automatic punishments which emanate from things themselves.' When told a story about a boy who steals an apple and then falls through a rotten bridge, the majority of six- to ten-year-olds say the bridge collapsed because the boy stole the apple. To the child, Piaget hypothesized, 'nature is a harmonious whole, obeying laws that are as much moral as physical.'

In secondary school, one of my classmates particularly adept at schoolyard psychological torture made a sport of teasing me about everything from the way I combed my hair to the way I laced my shoes. I wanted more than anything to kick him where it counted. But I never had to. On field day a wayward bicycle handlebar ripped open his scrotum. With his manhood maimed, I couldn't help but feel a sense of justice in the universe. He was asking for it. (Note that I do not actually advocate hitting below the belt. And don't worry – the boy was back at school the next day with stitches.)

We don't grow out of immanent-justice reasoning. One study found greater endorsement of karmic causality among university students than among eleven-year-olds. Another study found immanent-justice reasoning to be highly intuitive; we make more use of it when the cognitive resources normally used to suppress it are being taxed. And another study found that we also attribute *good* fortune to the deeds and character of the fortunate. Which explains the story of Gabriel Vivas:

In 2005, A Boeing 737 went down in the Peruvian jungle, killing thirty-seven people but sparing 'Gabby' Vivas and the five family members with him. Vivas, a father of five living in Brooklyn, New York, was known for acting as a neighbourhood peacekeeper, organizing youth baseball, and feeding neighbours who came over to use his pool. 'When you talk about miracles, here is a family that deserved one', a fellow parishioner told a reporter. One of his coworkers said, 'Gabby had so much good karma built up, it was enough to get his whole family out safe.' (Not enough, however, to keep him off the plane in the first place – or to keep it aloft and prevent him from witnessing the death of thirty-seven fellow passengers, all of whom presumably had 'bad karma'. But enough to 'miraculously' make him one of the, well, majority on the flight who survived.)

Immanent-justice reasoning can perhaps be explained by *just world theory*, pioneered by Melvin Lerner in the 1960s. Lerner argued that we have a desire to believe that the world is just, that people get what they deserve. Without a belief in fairness, you wouldn't bother with hard work, sacrifices, or good deeds – you'd never expect a payoff.

You'd also live in a state of anxiety, continuously bracing for unwarranted catastrophe. Lerner proposed nine main strategies for maintaining one's belief in a just world, all of which have now been demonstrated through research. The most studied are reinterpretation of the character and behaviour of a fate's recipient, together typically known as the blame-the-victim effect. (Kristina Olson has proposed a simpler mechanism than just-world reasoning for the blame-the-victim effect. It may simply be that the goodness or badness of an event 'rubs off' on anyone involved – she has found that children as young as three prefer fortunate people over unfortunate people.) Other tactics for maintaining justice beliefs include reinterpreting an outcome as more positive (or negative) to fit what we think someone deserves, working toward actual restitution or retribution, and holding faith that fate will balance things out in the end.

A recent study demonstrated the justice motive's relevance to everyday behaviour. Subjects who recalled three recent good breaks (finding a perfect parking spot, say) interpreted a self-relevant but ambiguous piece of information as bad news. They apparently had already surpassed their quota of good luck and expected something to come along and compensate for it. I see reactions like this all the time. A friend recently found $20 in a taxi. The driver first claimed it, then backtracked and said he didn't want it because something bad always happens to him when he finds money. Another friend recently found $60 at a party and immediately offered me a drink; apparently, he felt the windfall wasn't fully his to keep because he hadn't earned it. And when I saunter down to the subway and a train comes immediately I sometimes feel upset that my luck's been wasted; the next time, when I'll probably be in a rush, I'll have to wait forever.

While our mental tricks can lead to blaming victims or supporting the status quo (*the poor and powerless deserve their places in life*), they can also help us cope with calamity. In August 2002, Germany suffered the worst flood disaster in its history, a river overflow that destroyed twenty-five thousand houses, hospitals, and schools in the state of Saxony.

Psychologists surveyed 112 of the victims, many of whom said their lives had been in danger, and found that those with stronger just-world beliefs suffered less depression, anxiety, and paranoia in the flood's wake. And just-world beliefs enable coping with spinal-cord injuries, Janoff-Bulman found. It's unknown which justifying tactics any of these victims deployed – maybe they accepted their troubles as deserved and unavoidable, or they looked forward to a compensatory karmic payout – but presumably they found ways to make the world seem a little less wanton.

Although to believe in karma (in this loose, Western sense, not the original reincarnation sense) is not necessarily to infer a gavel-wielding god meting out punishments and rewards, it does require at least an expansive notion of Newton's third law of motion (every action has an equal and opposite reaction) in which moral force is registered alongside physical force. However, there is a nonmagical analogue of karma with very real dominion. In the world of human relations, what goes around typically does come around. People reciprocate kindness in kind. And word and deed travel far, manifesting in mysterious ways. So, as I like to picture myself as an old man telling my grandchildren, don't put shit on a boomerang.

Grasping at Straw Men

Magical thinking can serve as a sense-making strategy, and we rely on it more when we can't control or predict our environment. In chapter 3 we saw that fishermen on the open sea stick to superstitious rituals. In chapter 6 we saw that people read minds into cars and pets to explain erratic behaviour.

We're also more likely to see the hand of *fate* in times of uncertainty.

When we feel out of control we search for patterns in the world. Finding regularities allows us to plan our behaviour in a manageable environment. And in our search for order we often see order that isn't there.

Jennifer Whitson and Adam Galinsky have shown that when people feel out of control, they're more likely to see shapes in random noise, false correlations in financial reports, or conspiracies in strings of events. So the need for control can lead us to spot patterns (which may or may not exist), and, as explained earlier, pattern perception leads to agency detection; spooky coincidences and semblant conspiracies suggest creators. Such conspicuous alignments don't just happen on their own, right?

Our search for patterns can also lead us to link events to previous events as part of a transcendent moral order, leading to a belief in karma. And we connect events to later events, explaining them in terms of their consequences (this happened because it allowed that to happen), which is teleological reasoning.

Pointing to a karmic or teleological cause for an accident provides a comforting sense of understanding, but it's not a real understanding in that it offers no guidance in predicting similar events. A doctor would be laughed out of his profession if he published a medical paper arguing that infertility is caused by the opportunity it gives couples to find their destiny in adoption. (Our disgraced doctor might go on, however, to find his own true calling in theology, a field attuned to a different sort of sense making.)

So far we've considered three ways uncertainty might lead to belief in fate: we seek order in terms of karmic forces, teleological intervention, and meaningful patterns in general, each of which suggests an ordering agent. (These mechanisms may overlap.) But lack of personal control can also motivate us to *directly* hypothesize an external source of control, as opposed to drawing conclusions from evidence of order. Aaron Kay of Duke University has found that when people recall a recent incident over which they had no control, or when they unscramble sentences including words such as *chaotic*, *mayhem*, and *nonsense*, they become anxious. To allay their anxiety, they increase their estimates of the extent to which God or some other nonhuman entity controls events in the universe.

Believing someone else is in charge can reduce anxiety in several ways. First, it can increase a sense of personal control vicariously if we identify with the source of external control (the *my daddy could beat up your daddy* phenomenon). Second, trusting in an organized overlord also lets us believe that the world itself must be organized and thus predictable and navigable. This, too, increases a sense of personal control. A third way we gain a sense of personal control is by believing we can actively appeal to this overlord to act on our behalf, by saying the right prayers or obeying the right moral codes or making the right sacrifices. But Aaron Kay argues that it's not all about personal control. In fact, sometimes we eschew authority (as when the adoptive mother recoiled from the responsibility of deciding which little Indian girl to take home). Believing in external control can reduce anxiety in itself, he says. We just want the comfort of trusting that *someone* has his sober hands on the wheel.

Lead Foot

Of course there's danger in ceding too much control to your designated driver, particularly when he's imaginary. You could find yourself barrelling through a guardrail, thinking, *Gee, I would have taken the next exit instead of this cliff right here, but, hey, what am I gonna do?* Keeping your hands off the wheel, even when you don't like the direction life is taking, is called fatalism. You don't think your efforts will have any impact, so you sit on your palms. And sometimes wind up in a ravine.

It's unclear how easily one can change one's global beliefs about destiny, or how specifically one can tailor them to specific situations. But to the degree that one *can* nudge them around, it's worth looking at the consequences of fatalistic beliefs, both the magical and non-magical kind.

In chapter 5 we saw that a reduced sense of free will can make you dishonest, selfish, and lazy – a 'useless jerk', as I put it. Similarly,

decades of data have revealed that people with an *external locus of control*
– those who believe events are caused by other people, chance, fate, or
the nebulous powers that be – are less healthy, motivated, optimistic,
and successful than people who believe themselves to be in control of
their fates. And people who believe their intelligence or abilities to be
relatively fixed – a kind of personal destiny – give up in the face of
setbacks instead of working to improve. These threads reveal that
when we believe the internal and external factors influencing our fates
to be out of our control, we suffer.

Health researchers have defined fatalism in several ways, tapping
variously into popular notions of chance, God's will, and genetic determinism, but the pattern seems to be that fatalistic patients don't do so
well; they take illness as a death sentence. Several studies show that
fatalism is correlated with reduced cancer screening, even after controlling for socioeconomic status. Also, people who believe in predes-
tination don't see the benefits in a healthy diet or exercise. And
believing God is helping you recover from a heart attack actually does
help you recover, but believing he caused the heart attack slows you
down.

Fatalism also encourages the flouting of safety precautions. Belief
in destiny is positively correlated with dangerous driving, reduced
seatbelt use, and frequent unprotected sex. If it's your time to go, people figure, it's your time to go. They treat destiny as an agent with a
goal. A *secret* agent with a mark on your head. Dodge one bullet and it
will just take aim and try again. Some fishermen don't bother learning
to swim, seeing a dip in the drink as a sign that your time is up. Until
the 1980s, at some American ports you could even buy fishing boots
with lead in them to hasten your demise when you go overboard. *Oh,
you're driving me off a cliff? Let me help you with that accelerator.* Why
fight fate?

The *omission bias* can produce effects similar to fatalism. According to this bias, we prefer erring through inaction to erring through
action, as it produces less blame and regret. Therefore, we often have a

tendency to not act, making us shy about trying new things and altering the status quo. One study of blackjack players in Las Vegas – these are not people afraid of risk – found that 80 percent of their mistakes consisted of playing too passively rather than too aggressively, even though errors of omission were costlier than errors of commission. The bias can also prevent us from taking widely recommended action. People prefer not to vaccinate children if the vaccine has a small chance of killing the child, even if the chance of death without the vaccine is twice as high. Sometimes letting life take its course can be deadly.

People can learn to cope with pretty much anything, but that doesn't mean they have to. Those who give in to every outside force are buffeted about like a leaf on the wind. They take every challenge as a signal to give up or change direction. In 1998, a man named Mark Biancaniello broke the Guinness World Record in the noble sport of bee bearding by wearing no less than 87.5 pounds of bees on his person. I'm sure he was stung numerous times in the course of his training; I would have called it a day, but just think: would Biancaniello still be a household name if he were better at taking a hint?

Make Shit Happen

To say something was 'meant' to happen implies that its meaning is transcendent, its purpose a matter of fact. A superior being willed it to occur with an end in mind, and our job is to recognize and accept that end – to understand God's plan and submit to it, if not aid in its actualization.

What happens if you reject this notion? If you say that nothing, anywhere, was ever 'meant' to happen? You become an existentialist (specifically an atheistic existentialist, which is almost redundant). Existentialism is the ultimate 'shit happens' philosophy. Maybe your house burned down and something good came of it. But nothing good *had* to come of it, the house could easily have not burned down at all, and you could easily not have even been born in the first place. No one put you here and torched your house as a gift. Further, because nothing was ever 'meant' to happen, the biological organism you call your self was never 'meant' to behave any special way, so universal moral law goes right out the window. You're free to do what you want, as long as you're willing to experience the repercussions (including feelings of guilt).

Many find this freedom terrifying. The freedom to define right and wrong for yourself, the freedom to interpret life however you wish. Sartre wrote that 'man is condemned to be free: condemned, because he did not create himself, yet nonetheless free, because once cast into the world, he is responsible for everything he does'. Renouncing externally imposed goals and taboos increases our responsibility by widening our options. We have to think harder about what to want and do.

Research on the *choice overload effect* shows that in many situations increasing options can lead to worse well-being. Compared to people presented with six flavours of jam, those presented with twenty-four flavours find the decision-making process more difficult and frustrating, they're more dissatisfied with the jam they choose, and they're more likely to give up and not choose any jam at all. Anyone who's ever been single in a large city understands this predicament. Just as Sartre argues that we are 'condemned to be free', the psychologist Barry Schwartz has written about the 'tyranny of freedom'. He links the increased rate of depression over the last century to our unparalleled autonomy in choosing consumer products, careers, and mates. 'A good life may require constraints', he writes.

Even with only two alternatives – *should I stay or should I go?* – there can be trouble. Faced with a difficult decision and the potential for regret, or the requirement to justify our choice to others, we may procrastinate, or we may try to hand the responsibility to someone else. One place to pass the buck is upward, to the heavens. We might look for signs that tell us what we're meant to do. Or, if one of the options is to do nothing, we might just sit by and do that – obeying the omission bias and implicitly (or even explicitly) accepting that whatever happens was meant to happen. We punt and put it in the hands of fate.

Existentialism makes decisions harder in at least two ways. First, it erases any divine signs that might highlight or eliminate certain options. And second, it increases the possibility for regret, because you alone hold responsibility. But while existentialism can make life harder, it does not necessarily make life more empty. Denying any inherent meaning in the universe is not to deny meaning full stop. Each person creates his own meaning, decides his own values, defines his own destiny. Because existentialism is the ultimate 'shit happens' philosophy, it's also the ultimate 'do it yourself' philosophy. (Nihilism actually parallels existentialism on the 'shit happens' scale, but in its denial of even personally created meaning, it replaces 'do it yourself' with 'what's the point?') As George Bernard Shaw said, 'Life isn't about *finding* yourself; life is about *creating* yourself.' An obstacle in your path doesn't necessarily mean you're supposed to find yourself in a different direction. No, you're charged with deciding then and there whether to pick a new destination or hitch your britches and march over the obstacle. Freedom can feel imprisoning but ultimately it's empowering.

About a decade ago, when my amateur philosophy was even more amateurish, I told the woman I was dating that I was an existentialist because I was depressed and saw no inherent meaning in life. She told me I was *too* depressed to be an existentialist; I was fatalistic. She was right.

Growing up, one of my favourite books was *Alexander and the Terrible, Horrible, No Good, Very Bad Day*. Finally, a book that told it how it was. Alexander wakes up with gum in his hair, trips on a skateboard as he steps out of bed, and drops his jumper in the wash basin, all in the first sentence. And it doesn't let up. By the time I became acquainted with Holden Caulfield, Caulfield's thunder had been stolen; I'd found my sympathetic antihero.

Around first grade I decided Tuesday was my unlucky day. (My THNGVBD day.) Each week I woke up expecting a series of personal affronts, and I wasn't disappointed. It helps when you're looking for them, inviting them, twisting situations into them to confirm your hypothesis. I then added Thursdays to my shit list. Two unlucky days a week is a lot to bear, at any age. But still I added another day, and another. Eventually the whole week, every week, was my unlucky week. I just knew life was against me.

In my last year of school, I gave a talk to my fellow students about my battle with depression. Towards the end, I talked about one of the most important tactics I'd discovered in my fight for control over my life. I had come to realize that nothing that happens around us – no piece of news, no event – is inherently bad or good. It just is. In theory, we can interpret events however we wish. I was used to judging nearly everything as bad, and it was going to take a lot of work to change thinking habits that I'd been using my entire life. But instead of indulging in victimhood, I was beginning to fight these habits, allowing myself to believe that things could go my way.

Taking control of your interpretations is a very existential (and perhaps Zen) approach – one I was obviously still struggling with when my ex called me a fatalist, and one I will never perfect. I still have trouble finding a way to interpret burning my toast as an opportunity and not just a crisis, and I can't imagine what I would do if I ever lost a child. But the existential attitude that meaning is what you make it serves as the core of cognitive therapy, and it's a tactic worth practising in your everyday life.

If you find it difficult to put a positive spin on your situation, you can at least imagine how you could be worse off. Maybe it seems that this trick shouldn't improve your mood – now reality bites and alternate reality does too – but research on counterfactual thinking shows that it does make your actual reality seem better. Bronze medal winners are happier than silver medal winners, because instead of thinking about how they're one step down from endorsing shoes, they're thinking about how they're one step up from shining them. Okay, so you burned your toast. At least it's not your last piece of bread, and, when all's said and done, at least you're not impaled on a tree branch.

Negotiating with Fate

Even if you accept that all meaning is relative and subjective, it sure doesn't feel that way. Every time we create meaning, we project it back onto the world; we still attach value to the event. So if one counts the perception of absolute, objective meaning inherent in the world as a subtle form of magical thinking, then we can't escape magical thinking. But we can make it do our bidding, by bending those meanings to our own ends. You miss the bus; is that a Bad event? Have you no choice but to feel upset? No, quickly rewrite the script so that it's a Good event – maybe even a Purposefully Good event – because you get to enjoy the weather a bit longer or have an interesting conversation on the next bus. Then watch the clouds and have that conversation. With enough practice, cognitive restructuring becomes an iterative, almost fluid, give-and-take process where the best of all possible worlds seems to take shape around you as you move through it.

The adoptive mothers Klevan spoke with were good at controlling their interpretations: they ignored 'signs' they didn't like and acknowledged ones they did. 'One woman told me the most painful thing people said to her while she was going through infertility treatments was, "Maybe it's not meant to be"', Klevan says. 'It's interesting because it almost seems like the logical thing to say. What could be a greater sign from God that you're not supposed to be a parent than you can't make a baby?' But all the women explicitly rejected this notion.

Another woman had a 'shockingly strong' sense of self-determination, Klevan recalls. 'This was someone who made things happen.' The woman started having miscarriages and decided to adopt, but her husband was reluctant. 'She said, "Look, if you don't adopt with me, I'm leaving you because I want to be a mother NOW", and she went out and she made it happen.' But in telling her story, the woman noted a series of fortunate coincidences that paved the way for the adoption. 'The fact that she had worked really hard to make this happen did not mean that it wasn't also supposed to happen', Klevan says. 'It was like she was in perfect concert with the universe.'

'It seems to me that a lot of people have a sense of their own culpability woven into their ideas of fate', Michael Steger, a psychologist at Colorado State University, says. 'I think of a football player thanking God after scoring or winning. I don't get the sense that they think they just need to show up and bumble around to succeed. Rather, they need to meet the challenge that fate is laying before them. Kind of complex when you think about it, really.'

And successful people are often laying that fate out themselves and calling it a calling. When you feel like the universe is expecting your arrival at success, you're more prone to push through any difficulties and to see any suffering along the way as necessary and worthwhile. In negotiating the adoption process, 'you do have to believe that you're just fated to have a child', Cindy Bennett told me. 'Otherwise, you'll give up. Because it's not easy. It really isn't.'

A challenge to thinking in terms of fate is knowing when to recalibrate – when to let go of a lost cause and find a new fate to fulfil. If you try to define your own destiny with no regard for reality, you'll head off into la-la land where you're challenging the basketball great LeBron James to a one-on-one game, assured that the laws of gravity don't apply to you because it's your calling to show him who's king of the court. And when he wraps the court's basket around your head, you'll deludedly call it a crown. On the other hand, if you think that because you're five foot seven and have a funny name, like, say, Spud Webb, a career in basketball is not in the cards, you'll never win the 1986 Slam Dunk Contest with a 180-degree reverse two-handed strawberry jam (from a lob bounce off the floor). Finding the balance between steadfast and supple engagement comes from experience and is called wisdom. According to the gerontologist James Birren and the psychologist Laurel Fisher, 'Wisdom is tested by circumstances in which we have to decide what is changeable and what is not.' Then, once you've decided something isn't changeable, make peace with your decision. Don't give up before you need to, but once you do, commit to a new course and don't look back.

I like to employ a poi metaphor when thinking about when to go with the flow and when to fight the current. Poi is a form of dance in which one swings two small weights around at the ends of tethers, forming weaves and other dynamic patterns. Rule number one: respect the momentum of the poi. If you decide to abruptly yank your hand in a new direction while a poi is swinging around, you just might get a ball of fire in your face. (The poi are often soaked in fuel and lit up.) But you can bend their paths through the air, and with practice and finesse you can even decelerate them to a standstill and accelerate them on new paths in an instant. It's a constant negotiation between you and the poi. Similarly, many aspects of life have a kind of momentum – careers, relationships, conversations. If you inexpertly impose control and yank them on a new trajectory, you might have some fires to put out.

The poi, by the way, will inevitably go where you don't want them to at some point, perhaps ricocheting off your back or wrapping around your leg. If you react quickly, you just might make it look intentional and go on with the show. You may even have discovered a new trick. But the best trick of all is when you convince yourself that it was meant to happen.

Epilogue

The World Is Sacred

A Stab at a Secular Spirituality

At the outset of this book, I promised to convince even the most hard-core sceptics among you that you believe in magic. If I failed, I don't think the sword is ever coming out of that rock.

I also promised to demonstrate that magical thinking ain't so bad. It's a set of illusions, sure, but in many cases they're positive illusions. They primarily emerge from functions that have evolved for other reasons, but they have functionality of their own. First, magical thinking provides a sense of control. The value of an illusory sense of control is that it reduces anxiety and increases a feeling of agency, which can spur you to seize real control. Second, magical thinking provides meaning. There's meaning as in comprehension – understanding *how* things happen or *how* to do things – which allows for control. But there's also meaning as in a sense of purpose – grasping *why* things happen or *why* anything is worth doing. This is the stuff that gets you out of bed in the morning and lets you sleep at night.

While science should be your go-to guy for control and comprehension, you wouldn't bother trying to control or comprehend anything without a sense of something more.

Sacred Meaning

For a certain type of person, the question 'Is this all there is?' has a poignant, even urgent resonance. He goes about his day aware that nothing he does will ever save him from his ultimate fate of mental dissolution and physical decomposition. He's just an assortment of soulless atoms made to traverse a lifelong obstacle course towards a dead end. During his slog, tragedy strikes indiscriminately, joy is only temporary, and every little thing just seems like such a *hassle*. Seriously, *what's the point?*

This person is tied to his own insular individuality. He doesn't see anything beyond his downward spiral. But there's a way to break free of the cycle, at least for a moment: by experiencing the sacred.

A sense of sacredness dances in and out of the boundaries of magical thinking and touches on many of the same themes: essences, hidden forces, higher meanings. Scholars have had as difficult a time defining *sacredness* as they've had defining *religion* or *magic*, but they tend to agree that the sacred makes you feel connected to something larger than yourself. Sacredness gives religion its power, but it lives outside of religion, too. You might consider an object, a place, a time, a person, a value, or a role to be sacred. The presidency, an heirloom, honesty, an anniversary, the human body, the *Mona Lisa*, a favourite sports ground, freedom of speech, the Queen Mum. We cherish these things and see them as sources of communion and power that demand supreme respect. Sacred entities can't be reduced to their utilitarian functionality.

In Oscar Wilde's play *Lady Windermere's Fan*, Lord Darlington describes a cynic as someone 'who knows the price of everything and the value of nothing'. While cynics play an invaluable role in our society (they make great economists, it is said), you couldn't pay me enough to marry a full-time cynic. She'd be a robot, calculating the utility of everything and everyone around her. She'd lack both integrity and warmth. If, for her, objects and people and acts could not encapsulate broader abstractions – sentiments or identities or ideals – if she had no sense of a beyond, no capacity for awe or love or faith, then she would not be fully human. A person for whom nothing is sacred cannot be called one of us.

Certainly, holding things to be sacred can reduce flexibility and sharpen conflict. But it provides certain benefits. Kenneth Pargament, Annette Mahoney, and colleagues have found that sanctification of marriage is correlated with less marital strife and better collaborative problem solving. Sanctification of parenting is linked to less verbal aggression towards one's children. Sanctification of the environment is tied to greater preservation of the natural world. Sanctification of one's body is associated with more exercise, better diet, reduced drug and alcohol use, and improved body image. And sanctification of sex is linked to greater pleasure and satisfaction in the bedroom.

When one engages with the sacred, reality feels a little less flat. You get the vague sense that we're not just soulless atoms, that the world is imbued with a deep significance that can't be explained away. The answer to 'Is this all there is?' becomes a ringing 'No'. The ability to elicit the perception of sacredness on command would seem a valuable skill, and in fact you can train yourself in the art, according to one study. The psychologist Elisha Goldstein asked participants to practise mindful meditation for five minutes a day, five days a week, for three weeks. After each session, they turned their attention to what was special about a sacred or cherished object. For some it was a physical item such as a book, a plant, or a wedding ring. For others, it was a personal mantra or memory. Subjects reported feeling interconnection, gratitude, and peacefulness during their sacred moments. One man who saw the light explained, 'I never noticed any spiritual moments before this . . . [the words] *unique*, *holy*, and *worthy of reverence* [were] not within the scope of my intellectual reaction [to] things.' Goldstein found that at the end of the three weeks, and even six weeks after that, the subjects felt more satisfaction with life, felt greater purpose, and had more frequent spiritual experiences.

As I wrote in chapter 4, I'm a nonpeaker, so I decided to give the technique a try. Blip, my little red dragon, was gone, so as my sacred object I picked my silver jacket. It's a heavy aluminum-coated fire-retardant jacket I bought at a flea market fifteen years ago. It catalysed and emblemized my emerging self-confidence as a teenager, and I still use silverjacket.com as my domain name.

After trying the mindfulness technique every day for several weeks, I didn't have a single sacred moment. I think the problem is that I'm not yet adept at meditation. During this exercise, I was never able to keep my thoughts at bay for more than five consecutive seconds. But I didn't have high hopes that I would. My first year at Burning Man – the alternative music and arts festival in Nevada – I entered the 'Talk to God' art installation – a telephone booth standing in the middle of the desert with a sign that says TALK TO GOD – and asked the voice at the other end how I could live more in the moment. Its response: 'Breathe'. *Good idea*, I thought. I took a couple of slow breaths before wondering who was speaking at the other end of the line, whether anyone behind me was waiting to use the phone, and why I was no longer focused on breathing. But your mileage may vary – and presumably will depend on your motivation to practice.

I related my meditation struggles to Pelin Kesebir, a psychologist researching the perception of sacredness. 'Meditation might indeed not be everyone's cup of tea', she said. 'For such people it might work better to just be more mindful of opportunities to perceive sacredness in the world that pop up during the course of the day. Like when you're in nature, see a baby, or hear beautiful music.'

Kesebir has found that we often find sacredness in everyday life and, further, that finding sacredness in life acts as a buffer against existential threat. In one study, subjects more likely to consider various things sacred (the list included the national flag, animals, children, doctors, heirlooms, human life, motherhood, music, nature, and works of art) did not react to reminders of their mortality as strongly as their more cynical counterparts. They felt less threatened.

Great for those people, but what about the rest of us who don't reflexively see heaven in a wildflower? Kesebir found an easy exercise that can help. One group of subjects listed reasons water could be considered sacred (many noted its essential role in life). Another group simply listed practical uses for water. The first group showed reduced reactivity to reminders of death, compared to the second. Appreciating an everyday miracle reduced their existential angst. 'Perceptions of sacredness', she and her co-authors wrote, 'can instill a belief in the supreme value of life and the possibility of living meaningfully even in the face of ultimate annihilation.'

Sacred Motives

In *Man's Search for Meaning*, the psychiatrist Viktor Frankl describes his time as a death camp prisoner during the Holocaust. He emphasizes that to survive, prisoners needed a goal to live for. 'Being human always points, and is directed, to something, or someone, other than oneself – be it a meaning to fulfill or another human being to encounter', he wrote. 'Self-actualization is possible only as a side-effect of self-transcendence.'

The opposite of *sacred* is *profane*, but it could also be *banal*. 'Events that lack any relation to the higher, deeper, broader, or more integrative meanings of life can be called *banal*', the psychologist Roy Baumeister wrote in *Meanings of Life*. 'There is ample banality even in the most meaningful lives', he continued. 'Waiting in line, breathing, and urinating are good examples of acts that resist transcendental interpretations and are rarely a featured part of someone's life story.' But they don't have to be banal. If one imbues these everyday activities with purpose, they in a sense become sacred in that they make you feel as though you're contributing to a meaningful narrative, something larger than your in-the-moment flesh-and-blood self. I like to call this metaphysical upgrade the *wax on, wax off* approach.

In the film *The Karate Kid*, Mr Miyagi teaches karate to Daniel in exchange for manual labor. Daniel must wax Mr Miyagi's car, sand his deck, and paint his fence. The repetitive motions (wax on, wax off) are about as banal as breathing, until Daniel realizes he's been practising the proper mechanics of karate moves and uses the training to defeat his bullies in a local tournament. Suddenly, the monotonous chores go from menial to meaningful. If you could just treat all those unrewarding parts of your life – visiting the DFT, working a soul-crushing job, tying your shoelaces – as somehow edifying your character or enriching your life story, they'd gain just a hint of the sacred. Maybe you're not prepping to battle the Cobra Kai dojo at the All Valley Karate Tournament, but you can still find reward in perfecting a simple task, or noticing new details in the world, or thinking about the lives you're incrementally improving. Or maybe you simply transform the experience into a respectable exercise in endurance or self-discipline. Frankl argues that even a prisoner in pain facing certain death – about as meaningless a circumstance as one can imagine – has the opportunity to enrich his dwindling breaths: 'In accepting [the] challenge to suffer bravely, life has meaning up to the last moment.'

I believe the most meaningful lives include a balanced combination of long- and short-term goals. If you live for only the future and not the present, you'll never be happy, because at some point the future will be the present and you'll then be living for another future. But if you live for only the present, you'll severely impede your ability to enjoy the future because you'll be dealing with the fallout from today's decisions.

In the best of all worlds, the tasks you enjoy performing in the present will also align with more long-term pursuits. Tackling a series of short-term goals – on the order of minutes or seconds or fractions of a second – can lead to a state called *flow* where you feel fully engaged with an activity, neither bored nor anxious, and you lose your sense of self. But, coming out of that state, you want to feel it has advanced you as part of a larger mission. Playing video games may feel engrossing and satisfying in the moment but, on their own, would not make for a compelling autobiography.

||||||||||||||

On the importance of attaching present circumstances to future goals, Friedrich Nietzsche said, 'He who has a *why* to live can bear with almost any *how*.' I can attest to the power of meaningful destinations in giving life purpose. After signing a contract to write this book, I was able to go off antidepressants for the longest period in fifteen years. And research shows that people busy with even relatively rote activities are happier than people sitting idle. It's worth noting, however, that positive mood does not only follow meaning; you can't find meaning without being in the right mood. For instance, Laura King has shown that positive affect increases both magical thinking and a sense of purpose. And anyone who has taken antidepressants or stimulants can attest to the miraculous life-changing properties of these chemicals: the dopamine starts working, goals become desirable, and existence seems worth the hassle. Meaning 'automagically' springs from whatever you're doing.

Assuming you're in the right frame of mind (and chemistry of brain) to experience meaning, the meaning is there for you to create, whenever you want it. The historian of religion Mircea Eliade wrote in his classic book *The Sacred and the Profane*, 'every human experience is capable of being transfigured, lived on a different plane, a transhuman plane'. With a focus on living meaningfully even in minor tasks, your prosaic path becomes poetic; your mission gains a metaphysical significance. Every moment becomes sacred.

Breathe in, breathe out, wax on, wax off.

Sacred Miracles

Insane Clown Posse (ICP) is a rap duo from Detroit known for such albums as *Carnival of Carnage* and *Hell's Pit*. In April 2010, Violent J and Shaggy 2 Dope released a music video that caught many people off guard. The song, 'Miracles', is an earnest ode to all the wondrous, and some of the more mundane, elements in the world, from the Milky Way, childbirth, and rainbows down to water, air, fire, and dirt. The video is an uncanny and awe-inspiring sight to behold in itself: you have two guys with clown makeup and tattoos flying through space rapping about giraffes. Predictably, it drew instant ridicule, including a parody video called 'Magical Mysteries'. In response to the ICP lyric 'Fucking magnets, how do they work?' the parody's rappers asked inane questions like 'What the fuck is a clock?'

But the Posse was onto something. The rappers' mission wasn't to proclaim their ignorance about how the world works; it was to celebrate their wonder at its very existence. And the song wasn't necessarily about God, Violent J explained to an interviewer: 'A giraffe may not actually be a miracle according to the books, but I took my daughter and my son to the zoo last fall and we fed a giraffe. Let me tell you something – a giraffe is a fucking miracle. If you've never stood next to a fucking yellow-ass giraffe with a long neck, looking like a fucking dinosaur, it's just an amazing sight.' Even as unsentimental a character as the geek survivalist Dwight Schrute from the American version of *The Office* is on the same page as ICP; in one episode he is seen peeling magnets off the fridge and throwing them in the trash one by one. 'Garbage magnet. Garbage magnet', he says. 'God! Magnets are interesting enough; you don't need to tart 'em up with some design.' Einstein, too, had ICP's back when he said, 'There are two ways to live your life. One is as though nothing is a miracle. The other is as though everything is a miracle.'

One might think an existentialist would live his life the first way (no miracles), based on the premise that nothing was created for a purpose and everything is temporary; can anything incidental and ultimately inconsequential be called a miracle?

In a word, yes. Recall from chapter 5 that people reminded of their mortality found more meaning in life. (Perceived scarcity increases value, as any salesman who's ever said 'Available for a limited time only!' will tell you.) And while seeing fate working in your life can increase a sense of meaning (see chapter 7), feeling an extreme *lack* of fate can urge you to create your own meaning. Ronnie Janoff-Bulman wrote that trauma survivors, whose faith in a comprehensible and non-random world has been shaken, eventually shift 'from a concern with the meaning *of* life to a focus on meaning *in* life. . . . They become *committed* to living.' Since there's no ultimate point in your being here, there's no point in asking why you're here either. You just accept it and move on with the task of enjoying your stay.

The philosopher Irving Singer wrote of the positive side of existential anxiety:

> Once our hopeless questioning has reverberated in us, we may also intuit the mystery and wonder in everything being what it is. The source of our anxiety will not have changed but our attention will now be focused on the mere fact of existence rather than the obscure possibility of non-existence. We may also experience, at least occasionally, what Wittgenstein called 'astonishment that anything exists'. Instead of asking why there is something rather than nothing, we attend to the amazing – what may seem miraculous – presentation of any thing and every thing. . . . All reality will then appear to be what Santayana calls 'free entertainment'.

A clump of dirt, say, may not at first seem very amazing or entertaining, but it's free . . . and available for a limited time only.

|||||||||||||

Shaggy 2 Dope of ICP makes one false move when he raps, 'I don't wanna talk to a scientist / Y'all motherfuckers lyin', and gettin' me pissed'. But understanding the chemistry of water, or the physics of magnets, does not necessarily discount them as marvels. The physicist and Nobel laureate Richard Feynman once wrote, 'Poets say science takes away from the beauty of the stars – mere globs of gas atoms.

Nothing is "mere". I too can see the stars on a desert night, and feel them. But do I see less or more? . . . What men are poets who can speak of Jupiter if he were a man, but if he is an immense spinning sphere of methane and ammonia must be silent.'

So if nothing is 'mere', and yet (as I believe) nothing is a miracle in the supernatural sense, what else is there to say about reality but this: it is, pardon me, a *meracle*.

The Cosmological Constant

So things can be sacred because of their special histories (my silver jacket), or their symbolic significance (the act of coupling), or the simple fact of their existence (fucking magnets!). One could call attributions of sacredness a form of magical thinking, or not. In the end, I'm less interested in drawing definite boundaries around magical thinking than I am in what it does to us and for us.

The study of magical thinking is about more than silly sports superstitions or tribal voodoo hexes. It's much broader. It's about how we use symbols. It's about how we treat mysteries. It's about how we value things and people. It's about how we construct systems of ethics and justice. It's about how we think about life and death. It's about how we find reasons to go on.

Magical thinking guides how we define the very nature of the world, visible and invisible. Souls and luck and destiny and essences and mojo and free will and karma are more real, more alive in our daily thoughts and endeavours than quarks or quasars or quantum mechanics. They are intimate, ever present, and consequential parts of our cosmologies.

Magical thinking connects us simultaneously to the familiar, the unknown, and the unknowable. It allows us to see depth in the here and now, to form meaningful stories of our lives, to connect people and events in romantic and resonant ways. It also engages our imaginations about what forces swarm around us and what potentials lie within us. (Stories of wizards and superpowers are, at heart, not far-off fantasies but aspirational tales.) It presents a world full of untapped wonder, waiting for exploration and exploitation. Magical thinking both grounds us and keeps us buoyant.

Whether magic exists or not, magical thinking got us to where we are, and, for better or for worse, it will take us to where we're going. We could no sooner escape it than we could escape consciousness. We think, therefore we think magically.

|||||||||||||||

In struggling to unite my allegiance to critical thinking with my celebration of enchantment, I've no doubt spurned the loyalties of readers on both sides of the aisle. But whatever shaky synthesis I put forth will necessarily differ from anyone else's idea of a perfect balance. No amount of magical thinking is right for everyone.

And while writing this book I can't say that I've discovered the meaning of life – for myself, let alone for anyone else. But I have refined my thinking a bit about how to go about creating meaning. How to get out of my head, see the big picture, delight in the ineffable.

Without sacrificing scepticism, I've tempered my cynicism, and my world is a little more magical as a result.

Blip would be proud.

Acknowledgements

Thank you to all the scientists whose work contributed to this book directly or indirectly; the other sources whose insights and experiences enriched my perspective; my colleagues at *Psychology Today* magazine for providing support in the early stages of the book's gestation; friends and family for tolerating my extended cave dwelling; Susan Carnell, Carlin Flora, and Harry Hutson (aka Dad) for offering helpful feedback on the manuscript; Melissa Flashman (my excellent agent), Caroline Sutton (my excellent editor), and everyone on the Hudson Street Press team for making this book a reality; and my parents, for once upon a time creating something out of nothing.

NOTES

Introduction: We're All Believers

3 **'three human impulses':** Russell, B., *Why I Am Not a Christian: And Other Essays on Religion and Related Subjects* (New York: Simon and Schuster, 1957), 44.

5 **overestimate heights:** Jackson, R. E., and L. K. Cormack, 'Evolved Navigation Theory and the Descent Illusion,' *Perception and Psychophysics* 69 (2007): 353–62.

5 **overestimate sexual interest:** Haselton, M. G., and D. M. Buss, 'Error Management Theory: A New Perspective on Biases in Cross-Sex Mind Reading,' *Journal of Personality and Social Psychology* 78 (2000): 81–91.

5 **'predictably irrational':** Ariely, D., *Predictably Irrational: The Hidden Forces That Shape Our Decisions* (New York: HarperCollins, 2008).

5 **'*adaptively rational*':** Haselton, M. G., et al., 'Adaptive Rationality: An Evolutionary Perspective on Cognitive Bias,' *Social Cognition* 27 (2009): 737.

5 ***rational* system . . . *intuitive* system:** Epstein, S., 'Integration of the Cognitive and the Psychodynamic Unconscious,' *American Psychologist* 49 (1994): 709–24.

5 **we run largely on autopilot:** Bargh, J. A., and T. L. Chartrand, 'The Unbearable Automaticity of Being,' *American Psychologist* 54 (1994): 462–79.

6 **disconnects emotional brain centres:** Bechara, A. D., et al., 'Failure to Respond Autonomically to Anticipated Future Outcomes following Damage to Prefrontal Cortex,' *Cerebral Cortex* 6 (1996): 215–25.

6 **rationality or intelligence:** King, L. A., et al., 'Ghosts, UFOs, and Magic: Positive Affect and the Experiential System,' *Journal of Personality and Social Psychology* 92 (2007): 905–19; Irwin, H. J., *The Psychology of Paranormal Belief: A Researcher's Handbook* (Hatfield, UK: Univ. of Hertfordshire Press, 2009).

6 **'Magic – the very word':** Malinowski, B., *Magic, Science, Religion and Other Essays* (Glencoe, IL: Free Press, 1948), 50–51.

7 **'Although the word 'magic' ':** Nemeroff, C., and P. Rozin, 'The Makings of the Magical Mind,' in K. Rosengren, C. Johnson, and P. Harris (eds.), *Imagining the*

Impossible: Magical, Scientific, and Religious Thinking in Children (Cambridge, UK: Cambridge Univ. Press, 2000), 1.

7 **'confusion of subjectivity and objectivity':** Shweder, R., 'The Illusions of "Magical Thinking": Whose Chimera, Ours or Theirs?' (unpublished, 1994).

7 **'the anthropomorphism of nature':** Lévi-Strauss, C., *The Savage Mind* (Chicago: Univ. of Chicago Press, 1966), 221.

8 **'category mistakes':** Lindeman, M, and K. Aarnio, 'Superstitious, Magical, and Paranormal Beliefs: An Integrative Model,' *Journal of Research in Personality* 41 (2007): 734.

Chapter 1: Objects Carry Essences

14 **'copying machine':** Hood, B. M., and P. Bloom, 'Children Prefer Certain Original Objects over Perfect Duplicates,' *Cognition* 106 (2008): 455–62.

15 **'when a thing has history in it':** Dick, P. K., *The Man in the High Castle* (New York: Vintage, 1992; original work published 1962), 63–64.

15 **authentic objects:** Frazier, B. N., et al., 'Picasso Paintings, Moon Rocks, and Hand-Written Beatles Lyrics: Adults' Evaluations of Authentic Objects,' *Journal of Cognition and Culture* 9 (2009): 1–14.

16 **rubber ducky:** Frazier, B. N., and S. A. Gelman, 'Developmental Changes in Judgments of Authentic Objects,' *Cognitive Development* 24 (2009): 284–92.

16 *essentialism:* Gelman, S. A., *The Essential Child: Origins of Essentialism in Everyday Thought* (New York: Oxford Univ. Press, 2003).

16 **Mr Rogers:** Johnson, C. N., and M. G. Jacobs, 'Enchanted Objects: How Positive Connections Transform Thinking about the Very Nature of Things' (poster presented at the Society for Research in Child Development Biennial Meeting, Minneapolis, MN, April 2001).

16 **Richard Dawkins:** Barnes, R., and D. Hillman (producers), *The Genius of Charles Darwin* (London: IWC Media, 2008).

17 **December 1970:** BBC, 'George Michael Buys Lennon's Piano' (October 18, 2000), retrieved from http://news.bbc.co.uk/2/hi/entertainment/974485.stm.

17 **Jimmy Carter:** Turpin, C. (executive producer), *All Things Considered* [radio broadcast] (Washington, D.C.: National Public Radio, November 5, 2006).

17 **George Michael:** Zimmerman, A., 'Why Piano Owned by John Lennon Is Touring America,' *Wall Street Journal* (August 2, 2007), retrieved from http://online .wsj.com/public/article/SB118601703096585591.html.

19 **'secret sympathy':** Frazer, J. G., *The Golden Bough: A Study in Magic and Religion* (New York: Touchstone, 1995; original work published 1890), 14.

19 **rate various experiences:** Rozin, P., et al., 'Operation of the Laws of Sympathetic Magic in the Interpersonal Domain in American Culture,' *Bulletin of the Psychonomic Society* 27 (1989): 367–70.

20 **disease, misfortune, and evil:** Rozin, P., M. Markwith, and C. McCauley, 'Sensitivity to Indirect Contacts with Other Persons: AIDS Aversion as a Composite of Aversion to Strangers, Infection, Moral Taint, and Misfortune,' *Journal of Abnormal Psychology* 103 (1994): 495–505.

20 **'offensive to the taste':** Darwin, C., *The Expression of the Emotions in Man and Animals* (Chicago: Univ. of Chicago Press, 1965; original work published 1872), 256. Cited in Rozin, P., and A. E. Fallon, 'A Perspective on Disgust,' *Psychological Review* 94 (1987): 23–41.

20 **waste products:** Angyal, A., 'Disgust and Related Aversions,' *Journal of Abnormal and Social Psychology* 36 (1941): 393–412.

20 **mechanism has been co-opted:** Rozin, P., J. Haidt, and C. R. McCauley, 'Disgust,' in M. Lewis, J. M. Haviland-Jones, and L. F. Barrett (eds.), *Handbook of Emotions*, 3rd ed. (New York: Guilford Press, 2008), 757–76.

21 **disgusting film clip:** Horberg, E. J., et al., 'Disgust and the Moralization of Purity,' *Journal of Personality and Social Psychology* 97 (2009): 963–76.

21 **possibly through a *bad is gross* metaphor:** Royzman, E., and R. Kurzban, 'Minding the Metaphor: The Elusive Character of Moral Disgust,' *Emotion Review* 3 (2011): 269–71.

21 **jumper worn by their favourite celebrity:** Newman, G. E., G. Diesendruck, and P. Bloom, 'Celebrity Contagion and the Value of Objects,' *Journal of Consumer Research* (in press).

21 **evolutionary psychology:** Confer, J. C., et al., 'Evolutionary Psychology: Controversies, Questions, Prospects, and Limitations,' *American Psychologist* 65 (2010): 110–26.

22 **'Adolf Hitler':** Rozin et al., 'Operation of the Laws of Sympathetic Magic in the Interpersonal Domain in American Culture,' 369.

23 **properties of all kinds:** White, P. A., 'Property Transmission: An Explanatory Account of the Role of Similarity Information in Causal Interference,' *Psychological Bulletin* 135 (2009): 774–93.

23 **symbolism:** Nemeroff, C., and P. Rozin, 'The Contagion Concept in Adult Thinking in the United States: Transmission of Germs and of Interpersonal Influence,' *Ethos* 22 (1994): 158–86.

23 **eating fast-growing plants:** Meigs, A., *Food, Sex, and Pollution: A New Guinea Religion* (New Brunswick, NJ: Rutgers Univ. Press, 1984).

23 **eaten the hearts:** Frazer, *The Golden Bough*.

23 **wild boars and marine turtles:** Nemeroff, C., and P. Rozin, ''You Are What You Eat': Applying the Demand-Free 'Impressions' Technique to an Unacknowledged Belief,' *Ethos* 17 (1989): 50–69.

23 **vegetarianism:** Lindeman, M., P. Keskivaara, and M. Roschier, 'Assessment of Magical Beliefs about Food and Health,' *Journal of Health Psychology* 5 (2000): 195–209.

24 **Armin Meiwes:** Fager, J. (producer), *60 Minutes* (New York: CBS, March 16, 2008).

24 **'dying for a beer':** Sylvia, C., and C. Novak, *A Change of Heart: A Memoir* (New York: Grand Central Publishing, 1998), 110.

24 **'I seemed to inhale him':** Sylvia, C., 'I Was Given a Young Man's Heart,' *Daily Mail* (April 9, 2008), retrieved from http://www.dailymail.co.uk/health/article -558256/I-given-young-mans-heart---started-craving-beer-Kentucky-Fried -Chicken-My-daughter-said-I-walked-like-man.html.

25 **Israeli study:** Yoram, I., D. David, and I. Kutz, 'Another Person's Heart: Magical and Rational Thinking in the Psychological Adaptation to Heart Transplantation,' *Israel Journal of Psychiatry and Related Sciences* 41 (2004): 161–73.

25 **receiving tissue from a nonhuman:** Coffman, K., et al., 'Survey Results of Transplant Patients' Attitudes on Xenografting,' *Psychosomatics* 39 (1998): 379–83.

25 **donor of the same sex; essence:** Sanner, M. A., 'Transplant Recipients' Conceptions of Three Key Phenomena in Transplantation: The Organ Donation, the Organ Donor, and the Organ Transplant,' *Clinical Transplantation* 293 (2003): 391–400.

25 **'backwards causation':** Rozin et al., 'Operation of the Laws of Sympathetic Magic in the Interpersonal Domain,' 369.

26 **strings attached:** Truog, R. D., 'The Ethics of Organ Donation by Living Donors,' *New England Journal of Medicine* 353 (2005): 444–46.

26 *consubstantiation:* Miller, L. F., P. Rozin, and A. P. Fiske, 'Food Sharing and Feeding Another Person Suggest Intimacy: A Study of American College Students,' *European Journal for Social Psychology* 28 (1998): 423–36.

27 **Israeli soil:** Rozin, P., and S. Wolf, 'Attachment to Land: The Case of the Land of Israel for American and Israeli Jews and the Role of Contagion,' *Judgment and Decision Making* 3 (2008): 325–34.

27 **Yankee Stadium:** Schussler, R. M., 'Own a Piece of Yankee Stadium,' ABC News (September 19, 2008), retrieved from http://abcnews.go.com/Business/story?id =5835764.

28 **foot bath:** Chesterfield Football Club Official Website, 'Memorabilia Auction Raises £20,000,' June 17, 2010, retrieved from http://www.chesterfield-fc.co.uk/page/EndOfAnEra/0,,10435~2069998,00.html.

28 **T-shirt:** Argo, J. J., D. W. Dahl, and A. C. Morales, 'Positive Consumer Contamination: Responses to Attractive Others in a Retail Context,' *Journal of Marketing Research* 45 (2008): 690–701.

28 **pictures of germs:** Nemeroff, C., 'Magical Thinking about Illness Virulence: Conceptions of Germs from 'Safe' Versus 'Dangerous' Others,' *Health Psychology* 14 (1995): 147–51.

29 **Scarlett Johansson:** Leno, J. (producer), *The Tonight Show with Jay Leno* (New York: NBC, December 17, 2008).

29 **$5300:** BBC, 'Johansson Tissue Sells for $5300' (December 23, 2008), retrieved from http://news.bbc.co.uk/2/hi/entertainment/7796991.stm.

29 **infected with an STD:** Comer, L. K., and C. Nemeroff, 'Blurring Emotional Safety with Physical Safety in AIDS and STD Risk Estimations: The Casual/Regular Partner Distinction,' *Journal of Applied Social Psychology* 30 (2000): 2467–90.

29 **sexual violation:** Fairbrother, N., and S. Rachman, 'Feelings of Mental Pollution Subsequent to Sexual Assault,' *Behaviour Research and Therapy* 42 (2004): 173–89.

30 **changes in the way a victim thinks about the perpetrator:** Rachman, S., *Fear of Contamination: Illusions of Vulnerability* (Oxford: Oxford Univ. Press, 2006).

30 **Lady Macbeth effect:** Zhong, C., and K. Liljenquist, 'Washing Away Your Sins: Threatened Morality and Physical Cleansing,' *Science* 313 (2006): 145–46.

30 **Gamblers who washed their hands:** Xu, A. J., R. Zwick, and N. Schwarz, 'Washing Away Your (Good or Bad) Luck: Superstition, Embodiment, and Gambling Behavior,' *Journal of Experimental Psychology: General* (in press).

30 **'soak through all things':** James, W., *The Varieties of Religious Experience: A Study in Human Nature* (New York: Touchstone, 1997; original work published 1902), 61.

31 **Faecal-Vision glasses:** Michaels, L. (producer), *Saturday Night Live* (New York: NBC, March 16, 1996).

31 **kashrut:** Nemeroff and Rozin, 'The Makings of the Magical Mind.'

31 **stepping through someone's shadow:** C. Nemeroff, personal communication, August 14, 2007.

32 **Capgras delusion:** Ellis, H. D., and M. B. Lewis, 'Capgras Delusion: A Window on Face Recognition,' *Trends in Cognitive Sciences* 5 (2001): 149–56.

32 **buildings and household items:** Anderson, D. N., 'The Delusion of Inanimate Doubles: Implications for Understanding the Capgras Syndrome,' *British Journal of Psychiatry* 153 (1988): 694–99.

32 *SuperSense:* Hood, B. M., *SuperSense: Why We Believe in the Unbelievable* (San Francisco: HarperOne, 2009).

32 **spiritual-mystical states:** d'Aquili, E. G,. and A. B. Newberg, 'The Neuropsychology of Aesthetic, Spiritual and Mystical States,' *Zygon* 35 (2000): 39–52.

32 **Cotard delusion:** Ramachandran, V. S., *The Tell-Tale Brain: A Neuroscientist's Quest for What Makes Us Human* (New York: Random House, 2011). Also see McKay, R., and L. Cipolotti, 'Attributional Style in a Case of Cotard Delusion,' *Consciousness and Cognition* 16 (2007): 349–59.

33 ***mana:*** Stevens, P., 'Mana,' in *Encyclopedia of Anthropology* (Thousand Oaks, CA: Sage Publications, 2006), 1525–26.

33 **energy can have biological or mental attributes:** Svedholm, A. M., M. Lindeman, and J. Lipsanen, 'Believing in the Purpose of Events: Why Does It Occur, and Is It Supernatural?' *Applied Cognitive Psychology* 24 (2010): 252–65.

33 **California state assemblyman:** Brown, P. L., 'California Measure Would Align Building Rules with Feng Shui,' *New York Times* (January 30, 2004), retrieved from http://query.nytimes.com/gst/fullpage.html?res=9D03E6DA1138F933A05752C0 A9629C8B63.

33 **homeopathic remedies:** Stevens, P., 'Magical Thinking in Complementary and Alternative Medicine,' *Skeptical Inquirer* 25 (2001): 32–37.

33 **Energy healing:** National Center for Complementary and Alternative Medicine, 'What is Complementary and Alternative Medicine?' (April 1, 2011), retrieved from http://nccam.nih.gov/health/whatiscam/.

36 **golf putts:** Lee, C., et al., 'Putting Like a Pro: The Role of Positive Contagion in Golf Performance and Perception' (manuscript submitted for publication, 2010).

Chapter 2: Symbols Have Power

37 **Gold Dust Twins:** Beattie, J., 'Lynn Thrilled for Rice, His 'Gold Dust Twin' Hall of Famer,' *NESN* (July 23, 2009), retrieved from http://www.nesn.com/2009/07/lynn-thrilled-for-rice-his-gold-dust-twin-hall-of-famer.html.

38 **Curse of the Bambino:** Shaughnessy, D., *The Curse of the Bambino* (New York: Penguin, 2004).

38 **David Ortiz:** Shaughnessy, D., *Reversing the Curse* (New York: Houghton Mifflin Harcourt, 2005).

38 **Ortiz jersey:** Associated Press, 'Yankees Unearth Hidden Red Sox Jersey from New Stadium' (April 14, 2008), retrieved from http://sports.espn.go.com/mlb/news/story?id=3344825; Guarneri, B., M. Nestel, and A. Montefinise, 'Yankees' Hole of Fame,' *New York Post* (May 3, 2009), retrieved from http://www.nypost.com/p/news/regional/yankees_hole_of_fame_oVS9S58wE6FCZcJ2eAP0fL; Montefinise, A., et al., 'Hammering the Hex,' *New York Post* (April 13, 2008), retrieved from http://www.nypost.com/p/news/regional/item_iUPZZD0uk9z-kV9ugSUsjXI.

39 **'Memo to the *Post*':** New York Yankees, 'Statement from Alice McGillion, Spokesperson for the New York Yankees' [press release] (2008), retrieved from http://newyork.yankees.mlb.com/news/press_releases/press_release .jsp?ymd=20080412&content_id=2518720&vkey=pr_nyy&fext=.jsp&c_id=nyy.

39 **Hank Steinbrenner:** Montefinise et al., 'Hammering the Hex.'

40 **superstition online:** http://forums.nyyfans.com/showthread.php/110866-Red -Sox-Shirt-in-New-Stadium; http://gothamist.com/2008/04/12/proof_of_red_so.php; http://sports.espn.go.com/mlb/news/story?id=3344825; http://yankees.ipbhost.com/index.php?showtopic=14054.

40 **Yogi Berra:** Olshan, J., J. Nicholas, and C. Bennett, ''Under'Miner a Bx. Traitor,' *New York Post* (April 12, 2008), retrieved from http://www.nypost.com/p/news/regional/item_yK6HNCL6hqyt0fYZoymoUP.

40 **'The magician infers':** Frazer, *The Golden Bough*, 12.

41 'The principal key': Tylor, E. B., *Primitive Culture: Researches into the Development of Mythology, Philosophy, Religion, Language, Art, and Custom*, 6th ed., vol. 1 (London: John Murray, 1920; original work published 1874), 136.

41 JFK...darts: Rozin, P., L. Millman, and C. Nemeroff, 'Operation of the Laws of Sympathetic Magic in Disgust and Other Domains, *Journal of Personality and Social Psychology* 50 (1986): 703–12.

41 darts thrown at a photo of a baby: King et al., 'Ghosts, UFOs, and Magic: Positive Affect and the Experiential System,' *Journal of Personality and Social Psychology* 92 (2007): 905–19.

41 photography into parts of Africa: Behrend, H., 'Photo-Magic: Photographs in Practices of Healing and Harming in Kenya and Uganda,' *Journal of Religion in Africa* 33 (2003): 129–45.

42 distinctions between appearances and reality: Flavell, J. H., 'The Development of Children's Knowledge about the Appearance-Reality Distinction,' *American Psychologist* 41 (1986): 418–25.

42 Infants try to grasp pictures: DeLeoache, J., et al., 'Grasping the Nature of Pictures,' *Psychological Science* 9 (1998): 205–10.

42 picture of ice cream: Beilin, H., and E. G. Pearlman, 'Children's Iconic Realism: Object versus Property Realism,' in H. W. Reese (ed.), *Advances in Child Development and Behavior*, vol. 23. (New York: Academic Press, 1991), 73–111.

42 watching TV: Derrick, J. L., S. Gabriel, and K. Hugenberg, 'Social Surrogacy: How Favored Television Programs Provide the Experience of Belonging,' *Journal of Experimental Social Psychology* 45 (2009): 352–62.

42 fake turd and fake vomit: Rozin et al., 'Operation of the Laws of Sympathetic Magic in Disgust and Other Domains.'

42 alief: Gendler, T., 'Alief and Belief,' *Journal of Philosophy* 105 (2008): 634–63.

43 cutting up recognizable photos: Hood, B. M., et al., 'Implicit Voodoo: Electrodermal Activity Reveals a Susceptibility to Sympathetic Magic,' *Journal of Cognition and Culture* 10 (2010): 391–99.

43 rags for cleaning: Haidt, J., S. Koller, and M. Dias, 'Affect, Culture, and Morality, or Is It Wrong to Eat Your Dog?' *Journal of Personality and Social Psychology* 65 (1993): 613–28.

44 RadioShack: Orsi, P., 'Soccer: Voodoo Doll Promotion Spooks RadioShack,' Associated Press (January 29, 2009), retrieved from http://www.dispatch.com/live/content/sports/stories/2009/01/29/gameplan_voodoo.ART_ART_01-29-09_C2_9UCNMNM.html.

44 1976 presidential election: Carroll, J. S., 'The Effect of Imagining an Event on Expectations for the Event: An Interpretation in Terms of the Availability Heuristic,' *Journal of Experimental Social Psychology* 14 (1978): 88–96.

44 heuristics: Tversky, A., and D. Kahneman, 'Judgment under Uncertainty: Heuristics and Biases,' *Science* 185 (1974): 1124–31.

44 overestimate the danger: Lichtenstein, S., et al., 'Judged Frequency of Lethal Events,' *Journal of Experimental Psychology: Human Learning and Memory* 4 (1978): 551–78.

44 Craps and Magic: Henslin, J. M., 'Craps and Magic,' *American Journal of Sociology* 73 (1967): 316–30.

45 'natural prejudice': Mill, J. S., *A System of Logic, Ratiocinative and Inductive* (Honolulu: Univ. Press of the Pacific, 2002; original work published 1843), 533–34. Also see White, P. A., 'Property Transmission: An Explanatory Account of the Role of Similarity Information in Causal Inference,' *Psychological Bulletin* 135 (2009): 774–93.

45 pour water: Frazer, *The Golden Bough*.

45 **'Like Goes with Like':** Gilovich, T., and K. Savitsky, 'Like Goes with Like: The Role of Representativeness in Erroneous and Pseudo-Scientific Beliefs,' in T. Gilovich, D. W. Griffin, and D. Kahneman (eds.), *Heuristics and Biases: The Psychology of Intuitive Judgment* (Cambridge, UK: Cambridge Univ. Press, 2002), 617–24.

46 **graphology:** King, R. N., and D. J. Koehler, 'Illusory Correlation in Graphological Inference,' *Journal of Experimental Psychology: Applied* 6 (2000): 336–48.

46 **yellow fever:** Nisbett, R. E., and L. Ross, *Human Inference: Strategies and Shortcomings of Social Judgment* (Englewood Cliffs, NJ: Prentice-Hall, 1980).

46 **Doctrine of Signatures:** Pearce, J. M. S., 'The Doctrine of Signatures,' *European Neurology* 60 (2008): 51–52.

46 **pill colour:** de Craen, A. J. M., et al., 'Effect of Colour of Drugs: Systematic Review of Perceived Effect of Drugs and of Their Effectiveness,' *British Medical Journal* 313 (1996): 1624–26.

47 *Azzurri:* Moerman, D. E., *Meaning, Medicine, and the 'Placebo' Effect* (New York: Cambridge Univ. Press, 2002).

48 *metaphor therapy:* Li, X., L. Wei, and D. Soman, 'Sealing the Emotions Genie: The Effects of Physical Enclosure on Psychological Closure,' *Psychological Science* 21 (2010): 1047–50.

48 **Secular rituals:** Bell, C. M., *Ritual: Perspectives and Dimensions* (New York: Oxford Univ. Press, 1997).

50 *embodied cognition:* Barsalou, L. W., 'Grounded Cognition,' *Annual Review of Psychology* 59 (2008): 617–45.

50 **hot coffee:** Williams, L. E., and J. A. Bargh, 'Experiencing Physical Warmth Promotes Interpersonal Warmth,' *Science* 322 (2008): 606–7.

50–51 **clipboards . . . puzzle pieces:** Ackerman, J. M., C. C. Nocera, and J. A. Bargh, 'Incidental Haptic Sensations Influence Social Judgments and Decisions,' *Science* 328 (2010): 1712–15.

51 **complex metaphors:** Lakoff, G., and M. Johnson, *Philosophy in the Flesh: The Embodied Mind and Its Challenge to Western Thought* (New York: Basic Books, 1999).

51 **guitar players:** Fernandez, K. V., and J. L. Lastovicka, 'Making Magic: Fetishes in Contemporary Consumption,' *Journal of Consumer Research* (in press).

52 **other fetish objects:** Ellen, R., 'Fetishism,' *Man* 23 (1988): 213–35.

52 **devotional images . . . Qur'anic verses:** Behrend, 'Photo-Magic.'

52 *nominal realism:* Piaget, J., *The Child's Conception of the World*, J. Tomlinson and A. Tomlinson, trans. (Lanham, MD: Rowman and Littlefield Publishers, 2007; original work published 1929).

52 CYANIDE: Rozin et al., 'Operation of the Laws of Sympathetic Magic in Disgust and Other Domains.'

53 *lick, pick,* and *kick:* Hauk, O., I. Johnsrude, and F. Pulvermuller, 'Somatotopic Representation of Action Words in Human Motor and Premotor Cortex,' *Neuron* 41 (2004): 301–7.

53 NOT POISON: Rozin, P., M. Markwith, and B. Ross, 'The Sympathetic Magical Law of Similarity, Nominal Realism and Neglect of Negatives in Response to Negative Labels,' *Psychological Science* 1 (1990): 383–84.

53 *Britney:* http://www.ssa.gov/cgi-bin/babyname.cgi.

53 **'symbolic contamination of names':** Lieberson, S., *A Matter of Taste: How Names, Fashions, and Culture Change* (New Haven, CT: Yale Univ. Press, 2000), 131.

53 **share our . . . initials:** Pelham, B. W., M. Carvallo, and J. T. Jones, 'Implicit Egotism,' *Current Directions in Psychological Science* 14 (2005): 106–10.

54 **among employees:** Pilette, W. L., 'Magical Thinking by Inpatient Staff Members,' *Psychiatric Quarterly* 55 (1983): 272–74.

54 **'dirty words'**: Allan, K., and K. Burridge, *Forbidden Words: Taboo and the Censoring of Language* (New York: Cambridge Univ. Press, 2006), 242.

54 **thirteen:** Lachenmeyer, N., *13: The Story of the World's Most Notorious Superstition* (New York: Plume, 2004).

55 **Brussels Airlines:** Casert, R., '13 Dots in an Airplane Logo?' Associated Press (February 21, 2007), retrieved from http://www.usatoday.com/travel/flights/2007-02-21-brussels-airlines-superstitious-fliers-logo_x.htm.

55 **'I do':** Austin, J. L., *How to Do Things with Words*, 2nd ed. (Cambridge, MA: Harvard Univ. Press, 1975; original work published 1962), 5.

56 *induce motion in things':* Burke, K., *A Rhetoric of Motives* (Berkeley, CA: Univ. of California Press, 1969; original work published 1950), 42.

57 **'In records of anthropologists':** Cannon, W. B., ''Voodoo' Death,' *American Anthropologist* 44 (1942): 169.

57 **A 2002 retrospective:** Sternberg, E. M., 'Walter B. Cannon and ''Voodoo' Death': A Perspective from 60 Years On,' *American Journal of Public Health* 92 (2002): 1564–66.

57 **'Billy Goat':** Gatto, S., *Da Curse of the Billy Goat: The Chicago Cubs, Pennant Races, and Curses* (Lansing, MI: Protar House, 2004).

58 *stereotype threat:* For an overview, see Schmader, T., M. Johns, and C. Forbes, 'An Integrated Process Model of Stereotype Threat Effects on Performance,' *Psychological Review* 115 (2008): 336–56.

58 **'lovable losers':** Herrera, N., 'The Chicago Cubs and the Curse of a Stereotype (Part 1)' [blog post] (October 26, 2009), retrieved from http://www.psychologytoday.com/blog/personality-and-social-interaction/200910/the-chicago-cubs-and-the-curse-stereotype-part-1.

58 **Bartman:** Berkow, I., 'Memories of Bartman Die Hard for Cubs Fans,' *New York Times* (September 10, 2004), retrieved from http://www.nytimes.com/2004/09/10/sports/baseball/10bartman.html.

59 **Grant DePorter:** Altobelli, L., 'The Beat,' *Sports Illustrated* (February 28, 2005), retrieved from http://sportsillustrated.cnn.com/vault/article/magazine/MAG1109768/index.htm.

Chapter 3: Actions Have Distant Consequences

61 **deadliest job:** Bureau of Labor Statistics, 'Injuries, Illnesses, and Fatalities' (2010), retrieved from http://www.bls.gov/iif/oshcfoi1.htm.

63 **social transmission:** Higgins, S. T., E. K. Morris, and L. M. Johnson, 'Social Transmission of Superstitious Behavior in Preschool Children, *Psychological Record* 39 (1989): 307–23.

63 **hungry pigeons:** Skinner, B. F., ''Superstition' in the Pigeon,' *Journal of Experimental Psychology* 38 (1948): 168–72.

64 **foraging routines:** Timberlake, W., and G. A. Lucas, 'The Basis of Superstitious Behavior: Chance Contingency, Stimulus Substitution, or Appetitive Behavior?' *Journal of Experimental Analysis of Behavior* 44 (1985): 279–99.

64 **'About 5 min into the session':** Ono, K., 'Superstitious Behavior in Humans,' *Journal of the Experimental Analysis of Behavior* 47 (1987): 265.

66 *illusory correlation:* Shweder, R., 'Likeness and Likelihood in Everyday Thought: Magical Thinking in Judgments about Personality,' *Current Anthropology* 18 (1977): 637–48.

66 **near-hits:** Gilovich, T., 'Biased Evaluation and Persistence in Gambling,' *Journal of Personality and Social Psychology* 44 (1983): 1110–26.

67 **astrological personality profile:** Glick, P., and M. Snyder, 'Self-Fulfilling Prophecy: The Psychology of Belief in Astrology,' *Humanist* 50 (1986): 20–25.

67 **existence of ESP:** Russell, D., and W. H. Jones, 'When Superstition Fails: Reactions to Disconfirmation of Paranormal Beliefs,' *Personality and Social Psychology Bulletin* 6 (1980): 83–88.

67 *confirmation bias:* Nickerson, R. S., 'Confirmation Bias: A Ubiquitous Phenomenon in Many Guises,' *Review of General Psychology* 2 (1998): 175–220.

67 **Albert Michotte:** Michotte, A., *The Perception of Causality*, T. R. Miles and E. Miles, trans. (New York: Basic Books, 1963; original work published in 1945).

67 **consistency . . . not strictly required:** Woolley, J. D., C. A. Browne, and E. A. Boerger, 'Constraints on Children's Judgments of Magical Causality,' *Journal of Cognition and Development* 7 (2006): 253–77.

68 **illusion of control:** Langer, E., 'The Illusion of Control,' *Journal of Personality and Social Psychology* 32 (1975): 311–28.

68 **state lotteries:** Vyse, S. A., *Believing in Magic: The Psychology of Superstition* (New York: Oxford Univ. Press, 1997).

69 **'Either God is, or he is not'':** Pascal, B., *Pensées*, A. J. Krailsheimer, trans. (New York: Penguin, 1995; original work published 1669), 122.

69 **overestimate the romantic interest of women:** Haselton, M. G., and D. M. Buss, 'Error Management Theory: A New Perspective on Biases in Cross-Sex Mind Reading,' *Journal of Personality and Social Psychology* 78 (2000): 81–91.

69 **overestimate men's caddishness:** Geher, G., 'Accuracy and Oversexualization in Cross-Sex Mind-Reading: An Adaptationist Approach,' *Evolutionary Psychology* 7 (2009): 331–47.

69 **'paranoid optimists':** Haselton, M. G., and D. Nettle, 'The Paranoid Optimist: An Integrative Evolutionary Model of Cognitive Biases,' *Personality and Social Psychology Review* 10 (2006): 47–66.

69 *error management theory:* Haselton and Buss, 'Error Management Theory.'

69 **inflated self-esteem:** Taylor, S. E., and J. D. Brown, 'Illusion and Well-being: A Social Psychological Perspective on Mental Health,' *Psychological Bulletin* 103 (1988): 193–210.

70 **'whole system of principles':** Malinowski, B., *Magic, Science, Religion and Other Essays* (Glencoe, IL: Free Press, 1948), 13.

71 **New England fishermen:** Poggie, J. J., and R. B. Pollnac, 'Danger and Rituals of Avoidance among New England Fishermen,' *Maritime Anthropological Studies* 1 (1988): 66–78.

71 **Gulf War:** Keinan, G., 'Effects of Stress and Tolerance of Ambiguity on Magical Thinking,' *Journal of Personality and Social Psychology* 67 (1994): 48–58.

72 **'Malinowski Goes to College':** Rudski, J. M., and A. Edwards, 'Malinowski Goes to College: Factors Influencing Students' Use of Ritual and Superstition,' *Journal of General Psychology* 134 (2007): 389–403.

72 **'This is nonrational behavior':** Albas, D., and C. Albas, 'Modern Magic: The Case of Examinations,' *Sociological Quarterly* 30 (1989): 604.

73 **'As I listened to my professor':** Gmelch, G., 'Baseball Magic,' rev. ed., in E. Angeloni (ed.), *Annual Editions: Anthropology 11/12* (McGraw-Hill Contemporary Learning Series, 2010; essay originally published 1971), 150.

73 **'In professional baseball, fielding':** Ibid., 154.

73 **nursing home residents:** Langer, E. J., and J. Rodin, 'The Effects of Choice and Enhanced Personal Responsibility for the Aged: A Field Experiment in an Institutional Setting,' *Journal of Personality and Social Psychology* 34 (1976): 191–98; Rodin, J., and E. J. Langer, 'Long-Term Effects of a Control-Relevant Intervention with the Institutionalized Aged,' *Journal of Personality and Social Psychology* 35 (1977): 897–902.

74 **survival 'depended on one's ability':** Bettelheim, B., *The Informed Heart: Autonomy in a Mass Age* (Glencoe, IL: Free Press, 1960), 147.

74 **in-flight entertainment systems:** Charette, R. N., 'The Psychology of Comfortable Air Travel,' *IEEE Spectrum* (September 2008), retrieved from http://spectrum.ieee.org/consumer-electronics/gadgets/the-psychology-of-comfortable-air-travel.

75 **article about lifts:** Paumgarten, N., 'Up and Then Down: The Lives of Elevators,' *New Yorker* (April 21, 2008), retrieved from http://www.newyorker.com/reporting/2008/04/21/080421fa_fact_paumgarten.

75 **order of the tests:** Stotland, E., and A. Blumenthal, 'The Reduction of Anxiety as a Result of the Expectation of Making a Choice,' *Canadian Journal of Psychology* 18 (1964): 139–45.

75 **self-administered electric shocks:** Staub, E., B. Tursky, and G. E. Schwartz, 'Self-Control and Predictability: Their Effects on Reactions to Aversive Stimulation,' *Learning and Motivation* 18 (1971): 157–62.

75 **'even atheists may turn towards ritual practice':** Sosis, R., 'Psalms for Safety: Magico-Religious Responses to Threats of Terror,' *Current Anthropology* 48 (2007): 910.

76 *depressive realism:* Alloy, L. B., and L. Y. Abramson, 'Judgment of Contingency in Depressed and Nondepressed Students: Sadder but Wiser?' *Journal of Experimental Psychology: General* 108 (1979): 441–85; Msetfi, R. M., et al., 'Depressive Realism and Outcome Density Bias in Contingency Judgments: The Effect of the Context and Intertrial Interval,' *Journal of Experimental Psychology: General* 134 (2005): 10–22.

76 **learned helplessness:** Abramson, L. Y., M. E. P. Seligman, and J. D. Teasdale, 'Learned Helplessness in Humans: Critique and Reformulation,' *Journal of Abnormal Psychology* 87 (1978): 49–74.

77 **withholding negative feedback:** Matute, H., 'Learned Helplessness and Superstitious Behavior as Opposite Effects of Uncontrollable Reinforcement in Humans,' *Learning and Motivation* 25 (1994): 216–32.

78 **luck . . . as . . . attribute or a force:** Maltby, J., et al., 'Beliefs around Luck: Confirming the Empirical Conceptualization of Beliefs around Luck and the Development of the Darke and Freedman Beliefs around Luck Scale,' *Personality and Individual Differences* 45 (2008): 655–60; Wagenaar, W. A., and Keren, G. B., 'Chance and Luck Are Not the Same,' *Journal of Behavioral Decision Making* 1 (1988): 65–75.

78 **Maia Young:** Young, M. J., N. Chen, and M. W. Morris, 'Belief in Stable and Fleeting Luck and Achievement Motivation,' *Personality and Individual Differences* 47 (July 2009): 150–54.

78 **Michael Wohl and Michael Enzle:** Wohl, M. J. A., and M. E. Enzle, 'The Deployment of Personal Luck: Illusory Control in Games of Pure Chance,' *Personality and Social Psychology Bulletin* 28 (2002): 1388–97.

79 **Lysann Damisch:** Damisch, L., B. Stoberock, and T. Mussweiler, 'Keep Your Fingers Crossed!: How Superstition Improves Performance,' *Psychological Science* 21 (2010): 1014–20.

81 **Wade Boggs for eating chicken:** Curry, J., 'Boggs Is Beyond Compare, and He's the First to Say So,' *New York Times* (1994), retrieved from http://select.nytimes.com/gst/abstract.html?res=F1081EFA3B590C768DDDAA0894DC494D8.

81 **'embed superstitions in everyone's life':** Damisch, L., 'Keep Your Fingers Crossed!: The Influence of Superstition on Subsequent Task Performance and Its Mediating Mechanism' (Univ. of Cologne: unpublished doctoral dissertation, 2008), 85.

81 *30 Rock:* Wigfield, T., and J. Haller (writers), and D. Scardino (director), 'Future Husband' [television series episode], in T. Fey (creator), *30 Rock* (New York: NBC, March 11, 2010).

81 **Barack Obama played basketball:** Zeleny, J., 'No Ordinary Day for Obama: Back Home, a Bow to Superstition,' *New York Times* (November 5, 2008), retrieved from http://query.nytimes.com/gst/fullpage.html?res=9807E5DE1E3CF936A357 52C1A96E9C8B63.

81 **multiple-choice test:** Darke, P. R., and J. L. Freedman, 'Lucky Events and Beliefs in Luck: Paradoxical Effects on Confidence and Risk-Taking,' *Personality and Social Psychology Bulletin* 23 (1997): 378–88.

81 **London banks:** Fenton-O'Creevy, M., et al., 'Trading on Illusions: Unrealistic Perceptions of Control and Trading Performance,' *Journal of Occupational and Organizational Psychology* 76 (2003): 53–68.

82 **eclipses:** Lepori, G. M., 'Dark Omens in the Sky: Do Superstitious Beliefs Affect Investment Decisions?' (2009), retrieved from http://ssrn.com/abstract=1428792.

82 **'Don't be a fool':** Tambiah, S. J., *Magic, Science, Religion, and the Scope of Rationality* (Cambridge, U.K.: Cambridge Univ. Press, 1990), 54.

83 **children . . . magic rituals:** Evans, D. W., et al., 'Magical Beliefs and Rituals in Young Children,' *Child Psychiatry and Human Development* 33 (2002): 43–58.

83 **British boy:** Robertson, M. M., and A. E. Cavanna, 'The Disaster Was My Fault!' *Neurocase* 13 (2007): 446–51.

83 **OCD:** Einstein, D. A., and R. G. Menzies, 'The Presence of Magical Thinking in Obsessive-Compulsive Disorder,' *Behaviour Research and Therapy* 42 (2004): 539–49.

84 **exam takers:** Albas and Albas, 'Modern Magic.'

85 **exchanging a lottery ticket:** Risen, J. L., and T. Gilovich, 'Another Look at Why People Are Reluctant to Exchange Lottery Tickets,' *Journal of Personality and Social Psychology* 93 (2007): 12–22.

86 **applying to Stanford:** Risen, J. L., and T. Gilovich, 'Why People Are Reluctant to Tempt Fate,' *Journal of Personality and Social Psychology* 95 (2008): 293–307.

86 **negative scenarios:** Baumeister, R. F., et al., 'Bad Is Stronger Than Good,' *Review of General Psychology* 5 (2001): 323–70.

87 **get them called on:** Risen and Gilovich, 'Why People Are Reluctant to Tempt Fate.'

87 **newspaper articles:** Ibid.

87 **Anthrax:** Grove, L., 'The Reliable Source,' *Washington Post* (October 10, 2001), C3.

87 **'The universe seems interested':** Risen and Gilovich, 'Why People Are Reluctant to Tempt Fate,' 304.

88 **training doctors:** Naidech, A., N. K. Parek, and M. J. Kahn, 'Superstitions of House Officers,' *Resident and Staff Physician* 50 (2004): 47–49.

88 **Azerbaijan seclude infants:** Patterson, J., and A. Aghayeva, 'The Evil Eye: Staving off Harm – with a Visit to the Open Market,' *Azerbaijan International* 8 (2000): 55–57.

88 **gambling game:** Kruger, J., et al., 'Why Does Calling Attention to Success Seem to Invite Failure?' (paper presented at the Association for Consumer Research Conference, San Francisco, CA, November 2008).

89 **knocked on wood:** Keinan, G., 'The Effects of Stress and Desire for Control on Superstitious Behavior,' *Personality and Social Psychology Bulletin* 28 (2002): 102–8.

89 **'Refusing the added coverage':** Tykocinski, O. E., 'Insurance, Risk, and Magical Thinking,' *Personality and Social Psychology Bulletin* 34 (2008): 1355.

90 **'flouting conventional wisdom':** Risen and Gilovich, 'Why People Are Reluctant to Tempt Fate,' 303.

90 **anticipated regret:** Miller, D. T., and B. R. Taylor, 'Counterfactual Thought, Regret, and Superstition: How to Avoid Kicking Yourself,' in N. J. Roese and J. M. Olson (eds.), *What Might Have Been: The Social Psychology of Counterfactual Thinking* (Mahwah, NJ: Erlbaum, 1995), 305–32.

Chapter 4: The Mind Knows No Bounds

93 **the home crowd:** Moskowitz, T. J., and L. J. Wertheim, *Sportscasting: The Hidden Influences Behind How Sports Are Played and Games Are Won* (New York: Crown Archetype, 2011).

93 **Apollo 13:** Lovell, J. A., 'Houston, We've Had a Problem,' National Aeronautics and Space Administration (1975), retrieved from http://www.hq.nasa.gov/office/pao/History/SP-350/toc.html.

94 **induced nail-biting:** Arnold, M, 'Plight of 3 Apollo 13 Crewmen Stirs World Interest,' *New York Times* (April 15, 1970), retrieved from http://partners.nytimes.com/library/national/science/nasa/041570sci-nasa-arnold.html; Woodfill, J., 'Apollo 13: 'Houston, We've Got a Problem,'' National Aeronautics and Space Administration (1970), retrieved from http://er.jsc.nasa.gov/SEH/13index.htm; Woodfill, J., 'What Really Happened to Apollo 13' (August 19, 2009), retrieved from http://www.spaceacts.com/tract13.html.

95 **made a wish:** Woolley, J. D., et al., 'Where Theories of Mind Meet Magic: The Development of Children's Beliefs about Wishing,' *Child Development* 70 (1999): 571–87.

95 **Resisted certain thoughts:** Shafran, R., D. S. Thordarson, and S. Rachman, 'Thought-Action Fusion in Obsessive-Compulsive Disorder,' *Journal of Anxiety Disorders* 10 (1996): 379–91.

95 **Emily Pronin:** Pronin, E., et al., 'Everyday Magical Powers: The Role of Apparent Mental Causation in the Overestimation of Personal Influence,' *Journal of Personality and Social Psychology* 91 (2006): 218–31.

98 **'magnetically attract':** Byrne, R., *The Secret* (New York: Atria Books/Beyond Words), 10.

98 **'like placing an order':** Ibid., 48.

98 **Millions of people:** Chabris, C., and D. Simons, 'Fight 'The Power,'' *New York Times* (September 24, 2010), retrieved from http://www.nytimes.com/2010/09/26/books/review/Chabris-t.html.

99 **'primary cause of everything':** Byrne, *The Secret*, 33.

99 **'the things you can think . . . unlimited':** Ibid., 148.

99 **'events in history':** Ibid., 28.

99 **extraneous scientific-sounding information:** Weisberg, D. S., et al., 'The Seductive Allure of Neuroscience Explanations,' *Journal of Cognitive Neuroscience* 20 (2008): 470–77.

100 **Core to the scientific method:** Popper, K., *The Logic of Scientific Discovery*, 2nd ed. (New York: Routledge, 2002; original work published 1935).

101 **influence how we perceive:** Balcetis, E., and D. Dunning, 'See What You Want to See: Motivational Influences on Visual Perception,' *Journal of Personality and Social Psychology* 91 (2006): 612–25.

101 **look at a puzzle:** Wiseman, R., *The Luck Factor: Changing Your Luck, Changing Your Life: The Four Essential Principles* (New York: Hyperion, 2003).

102 **Expecting kindness:** Stinson, D. A., et al., 'Deconstructing the 'Reign of Error': Interpersonal Warmth Explains the Self-Fulfilling Prophecy of Anticipated Acceptance,' *Personality and Social Psychology Bulletin* 35 (2009): 1165–78.

102 **the severity of several illnesses:** Richman, L. S., et al., 'Positive Emotion and Health: Going Beyond the Negative,' *Health Psychology* 24 (2005): 422–29.

102 **optimism . . . boosts the immune system:** Segerstrom, S. C., and S. E. Sephton, 'Optimistic Expectancies and Cell-Mediated Immunity: The Role of Positive Affect,' *Psychological Science* 21 (2010): 448–55.

102 **health-enhancing behaviour:** Mulkana, S. S., and B. J. Hailey, 'The Role of Optimism in Health-Enhancing Behavior,' *American Journal of Health Behavior* 25 (2001): 388–95.

104 **'union of soul with body':** Hume, D., *An Enquiry concerning Human Understanding* (New York: Oxford Univ. Press, 2008; original work published 1748), 48.

104 **used an EEG:** Libet, B., et al., 'Time of Conscious Intention to Act in Relation to Onset of Cerebral Activity (Readiness-Potential): The Unconscious Initiation of a Freely Voluntary Act,' *Brain* 106 (1983): 623–42.

105 **electrodes in people's brains:** Fried, I., R. Mukamel, and G. Kreiman, 'Internally Generated Preactivation of Single Neurons in Human Medial Frontal Cortex Predicts Volition,' *Neuron* 69 (2011): 548–62.

105 **fMRI scanner:** Soon, C. S., et al., 'Unconscious Determinants of Free Decisions in the Human Brain,' *Nature Neuroscience* 11 (2008): 543–45.

106 **'We are enchanted by':** Wegner, D. M., 'Self Is Magic,' in J. Baer, J. C. Kaufman, and R. F. Baumeister (eds.), *Are We Free? Psychology and Free Will* (New York: Oxford Univ. Press, 2008), 226.

106 **'I'm a case in point':** Ibid., 237.

107 **'hard' problem of consciousness:** Chalmers, D. J., 'Facing Up to the Problem of Consciousness,' *Journal of Consciousness Studies* 2 (1995): 200–219.

108 **they were cheating:** Vohs, K. D., and J. Schooler, 'The Value of Believing in Free Will: Encouraging a Belief in Scientific Determinism Increases Cheating,' *Psychological Science* 19 (2008): 49–54.

108 **less willing to help:** Baumeister, R. F., E. J. Masicampo, and C. N. DeWall, 'Prosocial Benefits of Feeling Free: Disbelief in Free Will Increases Aggression and Reduces Helpfulness,' *Personality and Social Psychology Bulletin* 35 (2009): 260–68.

109 **day labour:** Stillman, T. F., et al., 'Personal Philosophy and Personnel Achievement: Belief in Free Will Predicts Better Job Performance,' *Social Psychological and Personality Science* 1 (2010): 43–50.

109 **watch . . . *Coffee and Cigarettes*:** Nordgren, L. F., et al., 'The Restraint Bias: How Illusions of Restraint Promote Impulsive Behavior,' *Psychological Science* 20 (2009): 1523–28.

110 **cup of coffee:** Williams and Bargh, 'Experiencing Physical Warmth.'

111 **'understand my enemy':** Card, O. S., *Ender's Game*, rev. ed. (New York: Tor, 1994; original work published 1985), 238.

112 **deciding appropriate prison sentences:** Carlsmith, K. M., J. M. Darley, and P. H. Robinson, 'Why Do We Punish? Deterrence and Just Deserts as Motives for Punishment,' *Journal of Personality and Social Psychology* 83 (2002): 284–99.

112 **'proportionate to their internal wickedness':** Kant, I., *The Science of Right*, (Charleston, SC: BibioBazaar, 2009; original work published 1790), 136.

112 **Shaun Nichols and Joshua Knobe:** Nichols, S., and J. Knobe, 'Moral Responsibility and Determinism: The Cognitive Science of Folk Intuitions,' *Noûs* 41 (2007): 663–85.

112 **'If retributivism runs that deep':** Greene, J. D., and J. D. Cohen, 'For the Law, Neuroscience Changes Nothing and Everything,' *Philosophical Transactions of the Royal Society of London B* 359 (2004): 1784.

113–14 **Apollo 14 . . . ESP experiment:** Mitchell, E., *The Way of the Explorer*, rev. ed. (Franklin Lakes, NJ: New Page Books, 2008; original work published 1996).

114 **Scientific studies endorsing . . . phenomena:** E.g., Bem, D. J., 'Feeling the Future: Experimental Evidence for Anomalous Retroactive Influences on Cognition and Affect,' *Journal of Personality and Social Psychology* 100 (2011): 1–19.

114 **scientists consider them unsound:** http://www.csicop.org/.

115 **'with a large enough sample':** Diaconis, P., and F. Mosteller, 'Methods for Studying Coincidences,' *Journal of the American Statistical Association* 84 (1989): 859.

116 **dreams foretell the future:** Morewedge, C. K, and M. I. Norton, 'When Dreaming Is Believing: The (Motivated) Interpretation of Dreams,' *Journal of Personality and Social Psychology* 96 (2009): 249–64.

116 **fortune cookie:** Lee, J.8, *The Fortune Cookie Chronicles: Adventures in the World of Chinese food* (New York: Twelve, 2008).

116 **orchestrated . . . coincidences:** Falk, R., 'Judgment of Coincidences: Mine versus Yours,' *American Journal of Psychology* 102 (1989): 477–93.

117 **'the skeptic asks':** Gilovich, T., *How We Know What Isn't So: The Fallibility of Human Reason in Everyday Life* (New York: Free Press, 1991), 180.

117 **Bruce Hood asked:** Hood, B. M., *SuperSense: Why We Believe in the Unbelievable* (San Francisco: HarperOne, 2009).

119 **'I was overwhelmed':** Mitchell, *The Way of the Explorer,* 75.

119 **'Songs, symphonies, movies':** Keltner, D., and J. Haidt, 'Approaching Awe, a Moral, Spiritual, and Aesthetic Emotion,' *Cognition and Emotion* 17 (2003): 310.

120 **'"oceanic"':** Freud, S., *Civilization and Its Discontents,* J. Strachey, trans. (New York: Norton, 1989; original work published 1930), 11.

120 **'The impulses to awe':** Dawkins, R., *Unweaving the Rainbow: Science, Delusion, and the Appetite for Wonder* (New York: Houghton Mifflin, 1998), 17.

120 **'supernatural' . . . 'numinous':** Hitchens, C., *God Is Not Great: How Religion Poisons Everything* (New York: Twelve, 2009), 286. Cited in Bloom, P., *How Pleasure Works: The New Science of Why We Like What We Like* (New York: Norton, 2010).

120 **wouldn't trust:** http://hitchensdebates.blogspot.com/.

120 **transcendence . . . magical thinking:** Neher, A., *The Psychology of Transcendence* (Mineola, NY: Dover, 1990; original work published 1980).

120 **nature . . . tyrannosaurus:** Shiota, M. N., D. Keltner, and A. Mossman, 'The Nature of Awe: Elicitors, Appraisals, and Effects on Self-Concept,' *Cognition and Emotion* 21 (2007): 944–63.

121–21 **purpose . . . committed and connected:** Saroglou, V., C. Buxant, and J. Tilquin, 'Positive Emotions as Leading to Religion and Spirituality,' *Journal of Positive Psychology* 3 (2008): 165–73.

121 **'The two religions':** Maslow, A. H., *Religions, Values, and Peak Experiences* (New York: Penguin, 1994; original work published 1964), 29.

121 **Apollo pilots:** Mitchell, *The Way of the Explorer.*

121 **[astronaut] memoirs:** Suedfeld, P., K. Legkaia, and J. Brcic, 'Changes in the Hierarchy of Value References Associated with Flying in Space,' *Journal of Personality* 78 (2010): 1411–46.

121 **'It is my strong suspicion':** Maslow, *Religions, Values, and Peak Experiences,* 75–76.

Chapter 5: The Soul Lives On

125 **Harry Fuller:** *The Argus,* 'Our Daughter's Killer Wrecked Our Lives Too' (March 26, 2004), retrieved from www.theargus.co.uk/archive/2004/3/26/116135.html; *Independent,* 'Couple's Killer Gets Two Life Sentences' (March 24, 1994), retrieved from http://www.independent.co.uk/news/uk/couples-killer-gets-two-life-sentences-security-video-and-tape-of-phone-call-trapped-insurance-man-1431145.html; R v Young [1995] QB 324 CA.

128 **electrical stimulation:** Selimbeyoglu, A., and J. Parvizi, 'Electrical Stimulation of the Human Brain: Perceptual and Behavioral Phenomena Reported in the Old and New Literature,' *Frontiers in Human Neuroscience* 4 (2010): 1–11.

128 **imagined biscuit:** Wellman, H. M., and D. Estes, 'Early Understanding of Mental Entities: A Reexamination of Childhood Realism,' *Child Development* 57 (1986): 910–23.

129 *Descartes' Baby*: Bloom, P., *Descartes' Baby: How the Science of Child Development Explains What Makes Us Human* (New York: Basic Books, 2004).

130 **'Baby Mouse'**: Bering, J. M., and D. F. Bjorklund, 'The Natural Emergence of Reasoning about the Afterlife as a Developmental Regularity,' *Developmental Psychology* 40 (2004): 217–33.

131 **'James Brown, move over'**: Preston, J., K. Gray, and D. M. Wegner, 'The Godfather of Soul,' *Behavioral and Brain Sciences* 29 (2006): 483.

132 **'Whenever we attempt to do so'**: Cited in Lifton, R. J., *The Broken Connection: On Death and the Continuity of Life* (New York: Simon and Schuster, 1979), 13.

132 **'Try, reader, to imagine'**: de Unamuno, M., *Tragic Sense of Life*, J. E. C. Fitch, trans. (New York: Dover, 1954; original work published 1913), 71.

133 **Richard Waverly**: Bering, J. M., 'Intuitive Conceptions of Dead Agents' Minds: The Natural Foundations of Afterlife Beliefs as Phenomenological Boundary,' *Journal of Cognition and Culture* 2 (2002): 263–308.

133 **'One particularly vehement extinctivist'**: Bering, J. M., *The Belief Instinct: The Psychology of Souls, Destiny, and the Meaning of Life* (New York: Norton, 2011), 118.

134 **Shaun Nichols**: Nichols, S., 'Imagination and Immortality: Thinking of Me,' *Synthese* 159 (2007): 215–33.

134 **Thomas Nagel**: Nagel, T., *The View from Nowhere* (New York: Oxford Univ. Press, 1989).

134 **Derek Parfit**: Parfit, D., *Reasons and Persons* (New York: Oxford Univ. Press, 1986).

135 **'we see death all around us'**: Nichols, 'Imagination and Immortality,' 230.

136 **Lord Horatio Nelson**: Hibbert, C., *Nelson: A Personal History* (Cambridge, MA: Perseus, 1994).

136 **'direct proof'**: Ramachandran, V. S., and W. Hirstein, 'The Perception of Phantom Limbs: The D. O. Hebb Lecture,' *Brain* 121 (1998): 1604.

136 **'Often, at night, I would try'**: Mitchell, S. W., 'The Case of George Dedlow,' *Atlantic Monthly* 18 (July 1866): 7.

136 **'UNITED STATES ARMY'**: Ibid., 11.

136 **donations and visitors**: Goler, R. I., 'Loss and the Persistence of Memory: 'The Case of George Dedlow' and Disabled Civil War Veterans,' *Literature and Medicine* 23 (2004): 160–83.

137 **'nearly every man who loses a limb'**: Mitchell, S. W., *Injuries of Nerves and Their Consequences* (Charleston, SC: BibioBazaar, 2009; original work published 1872), 348.

137 **somatosensory cortex**: Ramachandran and Hirstein, 'The Perception of Phantom Limbs.'

137 **'I stopped at the door to the room'**: Didion, J., *The Year of Magical Thinking* (New York: Knopf, 2005), 37.

138 **'*a person-file system*'**: Boyer, P., *Religion Explained: The Evolutionary Origins of Religious Thought* (New York: Basic Books, 2001), 219.

138 **'It keeps producing inferences'**: Ibid., 223.

138 **'offline social reasoning'**: Bering, J. M., 'The Folk Psychology of Souls,' *Behavioral and Brain Sciences* 29 (2006): 453–62.

138 **Even infants expect to find a person**: Jackson, E., J. J. Campos, and K. W. Fischer, 'The Question of Decalage between Object Permanence and Person Permanence,' *Developmental Psychology* 14 (1978): 1–10.

138 **'Out-of-sight, out-of-mind'**: Hodge, K. M., 'On Imagining the Afterlife,' *Journal of Cognition and Culture* (in press).

139 **'We should not be surprised'**: Boyer, *Religion Explained*, 227.

139 **'Just then, the wind chimes'**: Bering, J. M., 'The Cognitive Psychology of Belief in the Supernatural,' *American Scientist* 94 (2006): 148.

139 **Populus poll:** UK Polling Report, 'Populus Poll on Religion' (June 21, 2005), retrieved from http://ukpollingreport.co.uk/blog/archives/64.

139 **'really in touch':** GSS data, cited in MacDonald, W. L., 'Idionecrophanies: The Social Construction of Perceived Contact with the Dead,' *Journal for the Scientific Study of Religion* 31 (1992): 215–23.

139 **Most ghost stories:** McCue, P. A., 'Theories of Hauntings: A Critical Overview,' *Journal of the Society for Psychical Research* 66 (2002): 1–21.

140 **(debunked) five stages of grief:** Maciejewski, P. K., et al., 'An Empirical Examination of the Stage Theory of Grief,' *Journal of the American Medical Association* 297 (2007): 716–23.

140 **Jay Barham:** *Time*, 'The Conversion of Kübler-Ross' (November 12, 1979), retrieved from http://www.time.com/time/magazine/article/0,9171,946362-1,00.html.

140 **NDEs:** Blackmore, S., *Dying to Live: Near-Death Experiences* (Amherst, MA: Prometheus Books, 1993); Blanke, O., and S. Dieguez, 'Leaving Body and Life Behind: Out-of-Body and Near-Death Experiences,' in S. Laureys and G. Tononi (eds.), *The Neurology of Consciousness* (London: Academic Press, 2009), 303–25; Moody, R., *Life after Life* (San Francisco: HarperSanFrancisco, 2001; original work published 1975); Nelson, K. R., et al., 'Does the Arousal System Contribute to Near Death Experience?' *Neurology* 66 (2006): 1003–9.

141 **OBEs:** Blanke and Dieguez, 'Leaving Body and Life Behind'; Blanke, O., and G. Thut, 'Inducing Out-of-Body Experiences,' in S. D. Salla (ed.), *Tall Tales about the Mind and Brain: Separating Fact from Fiction* (New York: Oxford Univ. Press, 2007), 424–39; Lenggenhager, B., et al., 'Video Ergo Sum: Manipulating Bodily Self-Consciousness,' *Science* 317 (2007): 1096–99.

141 **message on a high shelf:** E.g., Horizon Research Foundation, 'An Interview with Dr. Sam Parnia on the Aware Study 2010' (2010), retrieved from http://www.horizonresearch.org/main_page.php?cat_id=233&pid=38.

142 **'seat of the soul':** Descartes, R., *The Passions of the Soul*, S. Voss, trans. (Indianapolis: Hacket Publishing Company, 1989; original work published 1649), 36.

142 ***push the argument:*** Wilson, N. L., and R. W. Gibbs, 'Real and Imagined Movement Primes Metaphor Comprehension,' *Cognitive Science* 31 (2007): 721–31.

143 **this man's soul weighed:** MacDougall, D., 'Hypothesis concerning Soul Substance Together with Experimental Evidence of the Existence of Such Substance,' *Journal of the American Society of Psychical Research* 1 (1907): 231–44.

143 **iconography:** Hodge, K. M., 'Descartes' Mistake: How Afterlife Beliefs Challenge the Assumption That Humans Are Intuitive Cartesian Substance Dualists,' *Journal of Cognition and Culture* 8 (2008): 387–415.

143 **Gene Roddenberry and Timothy Leary:** Simons, M., 'A Final Turn-on Lifts Timothy Leary Off,' *New York Times* (April 22, 1997), retrieved from http://www.nytimes.com/1997/04/22/world/a-final-turn-on-lifts-timothy-leary-off.html.

143 **cremains:** Kearl, M. C., 'Cremation: Purification, Desecration, or Convenience?' *Generations* 28 (2004): 15–20.

144 **Jon Stewart recounted:** Stewart, J. (host), *The Daily Show with Jon Stewart* (New York: Comedy Central, August 11, 2010).

144 **'I found to my horror':** Mitchell, 'The Case of George Dedlow,' 8.

145 **Columbine:** A Columbine Site, 'Basement Tapes' (n.d.), retrieved from http://acolumbinesite.com/quotes1.html; Cullen, D., *Columbine* (New York: Twelve, 2010); Zimmerman, 'Case Report No. 99-7625-A,' Jefferson County Sheriff's Office (1999), retrieved from http://www.davecullen.com/columbine/columbine-guide/video/basement-tapes-columbine.htm.

145 **Quigley:** Onion News Network, 'Scientists Successfully Teach Gorilla It Will Die Someday' [video file] (n.d.), retrieved from http://www.theonion.com/video/scientists-successfully-teach-gorilla-it-will-die,17165/.

146 'Mankind's common instinct': James, W., *The Varieties of Religious Experience: A Study in Human Nature* (New York: Touchstone, 1997; original work published 1902), 288.

147 'It doesn't matter': Becker, E., *The Denial of Death* (New York: Free Press, 1973), 5.

147 three early-career researchers: Solomon, S., J. Greenberg, and T. Pyszczynski, 'The Cultural Animal: Twenty Years of Terror Management Theory and Research,' in J. Greenberg, S. L. Koole, and T. Pyszczynski (eds.), *Handbook of Experimental Existential Psychology* (New York: The Guilford Press, 2004), 13–34.

148 Most TMT experiments: Burke, B. L., A. Martens, and E. H. Faucher, 'Two Decades of Terror Management Theory: A Meta-Analysis of Mortality Salience Research,' *Personality and Social Psychology Review* 14 (2010): 155–95.

148 worldview defense: For a review, see Solomon et al., 'The Cultural Animal.'

148 'immortality formula': Becker, *The Denial of Death*, 255.

149 self-esteem: For a review, see Pyszczynski, T., et al., 'Why Do People Need Self-Esteem? A Theoretical and Empirical Review,' *Psychological Bulletin* 130 (2004): 435–68.

149 abstract modern art: Landau, M. J., et al., 'Windows into Nothingness: Terror Management, Meaninglessness, and Negative Reactions to Modern Art,' *Journal of Personality and Social Psychology* 90 (2006): 879–92.

149 commitment to a romantic relationship: Florian, V., M. Mikulincer, and G. Hirschberger, 'The Anxiety Buffering Function of Close Relationships: Evidence That Relationship Commitment Acts as a Terror Management Mechanism,' *Journal of Personality and Social Psychology* 82 (2002): 527–42.

149 identification with their favourite brands: Rindfleisch, A., J. E. Burroughs, and N. Wong, 'The Safety of Objects: Materialism, Existential Insecurity, and Brand Connection,' *Journal of Consumer Research* 35 (2009): 1–16.

150 biosocial and creative: Lifton, *The Broken Connection*.

150 desire to have kids: Fritsche, I., et al., 'Mortality Salience and the Desire for Offspring,' *Journal of Experimental Social Psychology* 43 (2007): 753–62.

150 optimism about the football team: Dechesne, M., et al., 'Terror Management and the Vicissitudes of Sports Fan Affiliation: The Effects of Mortality Salience on Optimism and Fan Identification,' *European Journal of Social Psychology* 30 (2150): 813–35.

150 Famous people: Kesebir, P., and C.-Y. Chiu, 'The Stuff That Immortality Is Made Of: Existential Functions of Fame' (poster session presented at the annual meeting of the Society for Personality and Social Psychology, Albuquerque, NM, 2008).

151 'Every group': Rank, O., *Art and Artist: Creative Urge and Personality Development*, C. F. Atkinson, trans. (New York: Norton, 1989; original work published 1932), 411.

151 'These issues are germane': Lifton, *The Broken Connection*, 22.

152 bequests: Sargeant A., W. Wymer, and T. Hilton, 'Marketing Bequest Club Membership: An Exploratory Study of Legacy Pledgers,' *Nonprofit and Voluntary Sector Quarterly* 35 (2006): 384–404.

152 'Even existential philosophers': Lifton, *The Broken Connection*, 8.

152 the most direct evidence: Florian, V., and M. Mikulincer, 'Symbolic Immortality and the Management of the Terror of Death: The Moderating Role of Attachment Style,' *Journal of Personality and Social Psychology* 74 (1998): 725–34. Also see Dechesne, M., et al., 'Literal and Symbolic Immortality: The Effect of Evidence of Literal Immortality on Self-Esteem Striving in Response to Mortality Salience,' *Journal of Personality and Social Psychology* 84 (2003): 722–37.

153 Elvis: *Forbes*, 'The Top-Earning Dead Celebrities' (October 10, 2010), retrieved from http://www.forbes.com/2010/10/22/top-earning-dead-celebrities-business-entertainment-dead-celebs-10_land.html.

153 Quentin Tarantino: A Columbine Site, 'Basement Tapes.'

154 **sacrifice their individual selves:** Routledge, C., and J. Arndt, 'Self-Sacrifice as Self-Defense: Mortality Salience Increases Efforts to Affirm a Symbolic Immortal Self at the Expense of the Physical Self,' *European Journal of Social Psychology* 38 (2008): 531–41.

154 **Harris wrote in a journal:** Jefferson County Sheriff's Office, 'Columbine document JC-001-026007' (n.d.), retrieved from http://denver.rockymountainnews.com/pdf/900columbinedocs.pdf.

154 *optimal distinctiveness:* Simon, L., et al., 'Perceived Consensus, Uniqueness and Terror Management: Compensatory Responses to Threats to Inclusion and Distinctiveness Following Mortality Salience,' *Personality and Social Psychology Bulletin* 23 (1997): 1055–65.

154 **'we're gonna have followers':** A Columbine Site, 'Basement Tapes.'

155 **'martyrs like Eric and Dylan':** CNN, 'Shooter: 'You Have Blood on Your Hands'' (April 18, 2007), retrieved from http://edition.cnn.com/2007/US/04/18/vtech.nbc/.

155 **'mankind's real threat':** Becker, *The Denial of Death*, 227.

155 **'gods with anuses':** Ibid., 51.

155 **'Excreting is the curse':** Ibid., 33–34.

156 **bodily effluvia:** Cox, C. R., et al., 'Disgust, Creatureliness, and the Accessibility of Death-Related Thoughts,' *European Journal of Social Psychology* 37 (2007): 494–507.

157 **physical aspects of sex:** Goldenberg, J. L., et al., 'Understanding Human Ambivalence about Sex: The Effects of Stripping Sex of Its Meaning,' *Journal of Sex Research* 39 (2002): 310–20.

157 **foot binding and lip plates and neck rings:** Goldenberg, J. L., et al., 'I Am NOT an Animal: Mortality Salience, Disgust, and the Denial of Human Creatureliness,' *Journal of Experimental Psychology: General* 130 (2001): 427–35.

157 **'A woman is rendered':** de Beauvior, S., *The Second Sex*, H. M. Parshley, trans. (New York: Vintage, 1989; original work published 1949), 159.

158 **'The fetish takes "species meat"':** Becker, *The Denial of Death*, 236.

158 **'unmitigated testimonial':** Ibid., 237.

158 **'genital curses':** Stevens, P. 'Women's Aggressive Use of Genital Power in Africa,' *Transcultural Psychiatry* 43 (2006): 592–99.

158 **Jesus 'ate and drank':** Dunderberg, I. O., *Beyond Gnosticism: Myth, Lifestyle, and Society in the School of Valentinus* (New York: Columbia Univ. Press, 2008), 22.

158 **'I was once emptying':** Mather, C., *Diary of Cotton Mather, 1681–1724* (Boston: Massachusetts Historical Society, 1911), 357.

159 **'The God-like figure of Stalin':** Subbotsky, E., *Magic and the Mind: Mechanisms, Functions, and Development of Magical Thinking and Behavior* (New York: Oxford Univ. Press, 2010), 112.

160 **unhealthy ideals of beauty . . . :** For a review, see Goldenberg, J. L., 'The Body Stripped Down: An Existential Account of Ambivalence Toward the Physical Body,' *Current Directions in Psychological Science* 14 (2005): 224–28.

160 **not at the same time:** Although, Google 'blumpkin.'

160 *A Christmas Carol:* Dickens, C., *A Christmas Carol* (New York: Bantam, 1986; original work published 1843).

161 *the Scrooge effect:* Jonas, E., et al., 'The Scrooge Effect: Evidence That Mortality Salience Increases Prosocial Attitudes and Behavior,' *Personality and Social Psychology Bulletin* 28 (2002): 1342–53.

161 **'like the art of karate':** Kearl, M. C., 'Immortality, Symbolic,' in *Encyclopedia of Death and Dying* (n.d.), retrieved from http://www.deathreference.com/Ho-Ka/Immortality-Symbolic.html.

161 *Life's but a walking shadow:* Shakespeare, W., *Macbeth* (New York: Washington Square Press, 2003; original work published n.d.), 179.

161 *Fight Club:* Bell, R. G., C. Chaffin, and A. Linson (producers), Fincher, D. (director), *Fight Club* [motion picture] (United States: Regency Enterprises, 1999).

161 **'Although the *physicality* of death':** Yalom, I. D., *Existential Psychotherapy* (New York: Basic Books, 1980), 30.

161 **word-search puzzle:** King, L. A., J. A. Hicks, and J. Abdelkhalik, 'Death, Life, Scarcity, and Value: An Alternative Approach to the Meaning of Death,' *Psychological Science* 20 (2009): 1459–62.

162 **brushes with death:** Martin, L. L., W. K. Campbell, and C. D. Henry, 'The Roar of Awakening: Mortality Acknowledgment as a Call to Authentic Living,' in J. Greenberg, S. L. Koole, and T. Pyszczynski (eds.), *Handbook of Experimental Existential Psychology* (New York: Guilford Press, 2004), 431–38.

162 **'roar of awakening':** Kuhl, D., *What Dying People Want: Practical Wisdom for the End of Life* (New York: PublicAffairs, 2003), 227.

Chapter 6: The World Is Alive

165 **'We find human faces':** Hume, D., *Dialogues and Natural History of Religion* (New York: Oxford Univ. Press, 2009; original works published 1779, 1757), 141.

165 **'Hence the frequency and beauty':** Ibid., 141.

165 *Faces in the Clouds:* Guthrie, S. E., *Faces in the Clouds: A New Theory of Religion* (New York: Oxford Univ. Press, 1993).

165 **'the selecting power of nature':** Cited in ibid., 173.

165 **Piaget:** Piaget, J., *The Child's Conception of the World*, J. Tomlinson and A. Tomlinson, trans. (Lanham, MD: Rowman and Littlefield Publishers, 2007; original work published 1929).

166 **At the top of my field of vision:** Cited in Baron-Cohen, S., *Mindblindness: An Essay on Autism and Theory of Mind* (Cambridge, MA: MIT Press, 1995), 4–5.

167 **'echolocation':** Cited in ibid., 4.

167 **SEEK:** Epley, N., et al., 'On Seeing Human: A Three-Factor Theory of Anthropomorphism,' *Psychological Review* 114 (2008): 864–86.

167 **follow another's gaze:** Hood, B. M., J. D. Willen, and J. Driver, 'Adult's Eyes Trigger Shifts of Visual Attention in Human Infants,' *Psychological Science* 9 (1998): 131–34.

168 **communicative gestures:** Carpenter, M., K. Nagell, and M. Tomasello, 'Social Cognition, Joint Attention, and Communicative Competence from 9 to 15 Months of Age,' *Monographs of the Society for Research in Child Development* 63 (1998).

168 **recognize people's preferences:** Repacholi, B. M., and A. Gopnik, 'Early Reasoning about Desires: Evidence from 14- and 18-Month-Olds,' *Developmental Psychology* 33 (1997): 12–21.

168 **acknowledge goals:** Meltzoff, A. N., 'Understanding the Intentions of Others: Re-enactment of Intended Acts by 18-Month-Old Children,' *Developmental Psychology* 31 (1995): 838–50.

168 **discussing feelings:** Dunn, J., I. Bretherto, and P. Munn, 'Conversations about Feeling States between Mothers and Their Young Children,' *Developmental Psychology* 23 (1987): 132–39.

168 **false beliefs:** Baillargeon, R., R. M. Scott, and Z. He, 'False-Belief Understanding in Infants,' *Trends in Cognitive Sciences* 14 (2010): 110–18; Kovács, Á. M., E. Téglás, and A. D. Endress, 'The Social Sense: Susceptibly to Others' Beliefs in Human Infants and Adults,' *Science* 330 (2010): 1830–34.

168 **mirror neurons . . . emotions:** Carr, L., et al., 'Neural Mechanisms of Empathy in Humans: A Relay from Neural Systems for Imitation to Limbic Areas,' *Proceedings of the National Academy of Science* 100 (2003): 5497–5502.

168 **physical sensations:** Cheng, Y., et al., 'The Perception of Pain in Others Suppresses Somatosensory Oscillations: A Magnetoencephalography Study,' *NeuroImage* 40 (2008): 1833–40.

168 **intentions:** Iacoboni, M., et al., 'Grasping the Intentions of Others with One's Own Mirror Neuron System,' *PLoS Biology* 3 (2005): 529–35.

169 **egocentrism:** Epley, N., et al, 'Perspective Taking as Egocentric Anchoring and Adjustment,' *Journal of Personality and Social Psychology* 87 (2004): 327–39.

170 **'personality' in dogs:** Gosling, S. D., V. S. Y. Kwan, and O. P. John, 'A Dog's Got Personality: A Cross-Species Comparative Approach to Evaluating Personality Judgments,' *Journal of Personality and Social Psychology* 85 (2003): 1161–69.

170 **'Paro has internal states':** Wada, K., and T. Shibata, 'Living with Seal Robots: Its Sociopsychological and Physiological Influences on the Elderly at a Care House,' *IEEE Transactions on Robotics* 23 (2007): 974.

171 *primary emotions; Secondary emotions:* Demoulin, S., et al., 'Dimensions of Uniquely and Non-uniquely Human Emotions,' *Cognition and Emotion* 18 (2004): 71–96.

171 **'What the guilty look may be':** Horowitz, A., 'Disambiguating the 'Guilty Look': Salient Prompts to a Familiar Dog Behaviour,' *Behavioural Processes* 81 (2009): 451.

171 **74 percent of pet owners:** Morris, P., C. Coe, and E. Godsell, 'Secondary Emotions in Non-primate Species? Behavioral Reports and Subjective Claims by Animal Owners,' *Cognition and Emotion* 22 (2008): 3–20.

172 **mirror test:** For an overview, see de Waal, F. B. M., 'The Thief in the Mirror,' *PLoS Biology* 6 (2008): e201.

172 *episodic memory:* Episodic-*like* memory has been demonstrated in birds and rats. For a review, see Roberts, W. A., and M. C. Feeney, 'The Comparative Study of Mental Time Travel,' *Trends in Cognitive Sciences* 13 (2009): 271–77.

172 **'That tiger ain't go crazy':** Gallen, J., C. Rock, and M. Rotenberg (executive producers), *Chris Rock: Never Scared* [television broadcast] (New York: HBO, 2004).

173 **'Can they *suffer?*':** Bentham, J., *An Introduction to the Principles of Morals and Legislation* (New York: Oxford Univ. Press, 1995; original work published 1789), 283.

173 **natural environment:** Schultz, P. W., 'Empathizing with Nature: The Effects of Perspective-Taking on Concern for Environmental Issues,' *Journal of Social Issues* 56 (2000): 391–406.

173 **'If trees could scream':** Handey, J., *Deep Thoughts* (New York: Berkeley, 1992), 5.

174 **twenty sessions with AIBO:** Kanamori, M., M. Suzuki, and M. Tanaka, 'Maintenance and Improvement of Quality of Life among Elderly Patients Using a Pet-Type Robot,' *Japanese Journal of Geriatrics* 39 (2002): 214–18.

174 **chat rooms . . . AIBO's mental states:** Friedman, B., P. H. Kahn Jr., and J. Hagman, 'Hardware Companions? What Online AIBO Discussion Forums Reveal about the Human-Robotic Relationship,' *CHI Letters* 5 (2003): 273–80.

174 **survey of Roomba owners:** Sung, J.-A., et al., 'Housewives or Technophiles? Understanding Domestic Robot Owners,' *HRI 2008* (2008): 129–36.

174 **thirty Roomba owners:** Sung, J.-A., et al., ''My Roomba Is Rambo': Intimate Home Appliances,' *UbiComp 2007* (2007): 145–62.

174 **'when in *Star Wars*':** Gazzola, V., et al., 'The Anthropomorphic Brain: The Mirror Neuron System Responds to Human and Robotic Actions,' *NeuroImage* 35 (2007): 1683.

174 **human-computer interaction:** For a review, see Nass, C., and Y. Moon, 'Machines and Mindlessness: Social Responses to Computers,' *Journal of Social Issues* 56 (2000): 81–103.

175 'Ethiopians say that their gods': Lesher, J. H., *Xenophanes of Colophon: Fragments* (Toronto: Univ. of Toronto Press, 1992), 25.

175 intuitions don't always stick to scripture: Barrett, J. L., and F. C. Keil, 'Conceptualizing a Nonnatural Entity: Anthropomorphism in God Concepts,' *Cognitive Psychology* 31 (1996): 219–47.

176 'divine agents who can be influenced': Badcock, C., *The Imprinted Brain: How Genes Set the Balance of the Mind between Autism and Psychosis* (London: Jessica Kingsley Publishers, 2009), 120.

176 we read God's mind: Epley, N., et al., 'Believers' Estimates of God's Beliefs Are More Egocentric Than Estimates of Other People's Beliefs,' *Proceedings of the National Academy of Sciences* 106 (2009): 21533–38.

177 'That land that I live in': Dylan, B., 'With God on Our Side,' on *The Times They Are a-Changing'* [album] (New York: Columbia, 1964).

177 'The denial of determinate': Cited in Guthrie, *Faces in the Clouds*, 183.

178 two dimensions: Gray, H. M., K. Gray, and D. M. Wegner, 'Dimensions of Mind Perception,' *Science* 315 (2007): 619.

178 black-and-white animation: Heider, F., and M. Simmel, 'An Experimental Study of Apparent Behavior,' *American Journal of Psychology* 57 (1944): 243–59.

179 movement cues: For reviews, see Rakison, D. H., and D. Poulin-Dubois, 'Developmental Origin of the Animate-Inanimate Distinction,' *Psychological Bulletin* 127 (2001): 209–28; Scholl, B. J., and P. D. Tremoulet, 'Perceptual Causality and Animacy,' *Trends in Cognitive Sciences* 4 (2000): 299–309.

179 'Banging our eye': Guthrie, *Faces in the Clouds*, 47.

179 Infants as young as twelve months: Kuhlmeier, V., K. Wynn, and P. Bloom, 'Attribution of Dispositional States by 12-Month-Olds,' *Psychological Science* 14 (2003): 402–8.

180 Neuroimaging: Castelli, F., et al., 'Movement and Mind: A Functional Imaging Study of Perception and Interpretation of Complex Intentional Movement Patterns,' *NeuroImage* 12 (2000): 314–25.

180 'haunted scrotum': Harding, J. R., 'The Case of the Haunted Scrotum,' *Journal of the Royal Society of Medicine* 89 (1996): 600.

180 cheese toastie: Associated Press, '"Virgin Mary Grilled Cheese' Sells for $28,000'' (November 23, 2004), retrieved from http://www.msnbc.msn.com/id/6511148/.

180 rock formation on Mars: Phillips, T, 'Unmasking the Face on Mars,' *Science@ NASA Headline News* (May 24, 2001), retrieved from http://science.nasa.gov/science-news/science-at-nasa/2001/ast24may_1/.

180 shows a demon: Snopes, 'Faces in the Cloud' (April 23, 2008), retrieved from http://www.snopes.com/rumors/wtcface.asp.

180 facelike objects: Hadjikhani, N., et al., 'Early (N170) Activation of Face-Specific Cortex by Face-Like Objects,' *NeuroReport* 20 (2009): 403–7.

181 robotic faces: Dubal, S., et al., 'Human Brain Spots Emotion in Non Humanoid Robots,' *Social Cognitive and Affective Neuroscience* 6 (2011): 90–97.

181 smiley faces: Eger, E., et al., 'Rapid Extraction of Emotional Expression: Evidence from Evoked Potential Fields during Brief Presentation of Face Stimuli,' *Neuropsychologia* 41 (2003): 808–17.

181 brown furry lump: Johnson, S., V. Slaughter, and S. Carey, 'Whose Gaze Will Infants Follow? The Elicitation of Gaze Following in 12-Month-Olds,' *Developmental Science* 1 (1998): 233–38.

181 'uncanny valley': MacDorman, K. F., et al., 'Too Real for Comfort: Uncanny Responses to Computer Generated Faces,' *Computers in Human Behavior* 25 (2009): 695–710.

181 dog and cat prototypes: Shibata, T., and K. Wada, 'Robot Therapy: A New Approach for Mental Healthcare of the Elderly,' *Gerontology* 57 (2011): 378–86.

182 **foetus appeared to be smiling:** Borland, S., 'The Foetus Who Broke into a Big Smile . . . Aged Only 17 Weeks,' *Daily Mail* (October 11, 2010), retrieved from http://www.dailymail.co.uk/health/article-1319373/The-foetus-broke-big-smile --aged-17-weeks.html.

182 *LOLFetus:* O'Connor, M., 'Smiling Fetus Joins Abortion Debate' [blog post], (October 11, 2010), retrieved from http://gawker.com/5661238/smiling-fetus -joins-abortion-debate.

182 **statue of the athlete Nikon:** Evans, E. P., *The Criminal Prosecution and Capital Punishment of Animals* (London: William Heinemann, 1906).

182 **Mexico City . . . bell:** Cancino, F., 'Restauran el reloj de la Catedral tras nueve años,' *El Universal* (July 21, 2006), retrieved from http://www2.eluniversal.com .mx/pls/impreso/version_imprimir.html?id_nota=77995&tabla=ciudad.

183 **dyadic template of morality:** Gray, K., and D. M. Wegner, 'Blaming God for Our Pain: Human Suffering and the Divine Mind,' *Personality and Social Psychology Review* 14 (2010): 7–16.

183 **predators and prey:** New, J., .L. Cosmides, and J. Tooby, 'Category-Specific Attention for Animals Reflects Ancestral Priorities, Not Expertise,' *Proceedings of the National Academy of Sciences* 104 (2007): 16593–16603.

184 **'It is better for a hiker':** Guthrie, *Faces in the Clouds*, 6.

184 **Fear further biases:** Epley, N., et al., 'Creating Social Connection through Infer-ential Reproduction: Loneliness and Perceived Agency in Gadgets, Gods, and Greyhounds,' *Psychological Science* 19 (2008): 114–20.

184 **predominant view:** E.g., Boyer, P., *Religion Explained: The Evolutionary Origins of Religious Thought* (New York: Basic Books, 2011).

184 **Jesse Bering, Dominic Johnson:** E.g., Bering, J. M., and D. D. P. Johnson, ''O Lord . . . You Perceive My Thoughts from Afar': Recursiveness in the Cognitive Evo-lution of Supernatural Agency,' *Journal of Cognition and Culture* 5 (2005): 118–42.

184 **a bit like feathers:** Gould, S. J., 'Exaptation: A Crucial Tool for Evolutionary Psychology,' *Journal of Social Issues* 47 (1991): 43–65.

185 **subtle reminder of God:** Shariff, A. F., and A. Norenzayan, 'God Is Watching You: Supernatural Agent Concepts Increase Prosocial Behavior in an Anonymous Economic Game,' *Psychological Science* 18 (2007): 803–9.

185 **'Our character is what we do':** Brown, H. J., Jr., *P.S. I Love You: When Mom Wrote She Always Saved the Best for Last* (Nashville: Routledge Hill Press, 1990), 125.

185 **'Conscience is the inner voice':** Mencken, H. L., *A Mencken Chrestomathy* (New York: Vintage, 1982; original work published 1949), 617.

185 **exposure to . . . 'police' and 'jury':** Shariff and Norenzayan, 'God Is Watching You.'

185 **honesty box:** Bateson, M., D. Nettle, and G. Roberts, 'Cues of Being Watched Enhance Cooperation in a Real-World Setting,' *Biology Letters* 2 (2006): 412–14.

186 **internalize social norms:** Gintis, H., 'The Hitchhiker's Guide to Altruism: Gene-Culture Coevolution and the Internalization of Norms,' *Journal of Theoreti-cal Biology* 220 (2003): 407–18.

186 **the world is just:** Lerner, M. J., 'The Justice Motive: Some Hypotheses as to Its Origins and Forms,' *Journal of Personality* 45 (1977): 1–32.

186 **capacity for empathy:** Zak, P. J., A. A. Stanton, and S. Ahmadi, 'Oxytocin Increases Generosity in Humans,' *PLoS ONE* 2 (2007): e1128.

186 **'Jumanji':** Harwood, J., 'Is Dull World Cup Down to the 'Terrible' Jabulani Ball?' *First Post* (June 17, 2010), retrieved from http://www.thefirstpost.co.uk/64657,spo rt,football,terrible-jabulani-ball-blamed-for-dull-world-cup.

186 **'Obviously it's quite unpredictable':** BBC, 'World Cup 2010: David James Criticises Jabulani Ball' (June 2, 2010), retrieved from http://news.bbc.co.uk/sport2/hi/football/world_cup_2010/8716699.stm.

186 **'It's very weird':** Associated Press, 'World Cup Players Complain about 'Terrible' Adidas Ball' (May 31, 2010), retrieved from http://www.nesn.com/2010/05/world-cup.html.

186 **'the jackpot':** Guthrie, *Faces in the Clouds*, 45.

187 *effectance motivation:* White, R. W., 'Motivation Reconsidered: The Concept of Competence,' *Psychological Review* 66 (1959): 297–332.

187 **'If everywhere in nature':** Freud, S., *The Future of an Illusion*, J. Strachey, trans. (New York: Norton, 1989; original work published 1927), 20–21.

187 **unpredictable dog:** Epley, N., et al., 'When We Need a Human: Motivational Determinants of Anthropomorphism,' *Social Cognition* 26 (2008): 143–55.

187 **series of experiments:** Waytz, A., et al., 'Making Sense by Making Sentient: Unpredictability Increases Anthropomorphism,' *Journal of Personality and Social Psychology* 99 (2010): 410–35.

188 **'not be mere metaphor':** Ibid., 419.

188 **intentional stance:** Dennett, D. C., *The Intentional Stance* (Cambridge, MA: MIT Press, 1987).

189 **ancient navigators:** Guthrie, *Faces in the Clouds*.

189 **loneliness:** Epley et al., 'Creating Social Connection through Inferential Reproduction.'

190 **[pets' and] Paro's therapeutic value:** For a review, see Shibata and Wada, 'Robot Therapy.'

191 **everything anyone does as intentional:** Rosset, E., 'It's No Accident: Our Bias for Intentional Explanations,' *Cognition* 108 (2008): 771–80; Bègue, L., et al., 'There Is No Such Thing as an Accident, Especially When People Are Drunk,' *Personality and Social Psychology Bulletin* 36 (2010): 1301–4.

192 **legally required to undergo an ultrasound:** Sanger, C., 'Seeing and Believing: Mandatory Ultrasound and the Path to a Protected Choice,' *UCLA Law Review* 56 (2008): 351–408.

193 **'the same rights':** Waytz, A., N. Epley, and J. T. Cacioppo, 'Social Cognition Unbound: Psychological Insights into Anthropomorphism and Dehumanization,' *Current Directions in Psychological Science* 19 (2010): 61.

Chapter 7: Everything Happens for a Reason

197 **'I don't see why':** Hemingway, E., *The Snows of Kilimanjaro and Other Stories*, (New York: Scribner, 1995; original work published 1961; story first published 1936), 6.

198 **'That it should collapse':** Evans-Pritchard, E. E., *Witchcraft, Oracles and Magic among the Azande* (Oxford: Clarendon Press, 1963; original work published 1937), 69–70.

198 **intervening through natural means:** Legare, C. H., and S. A. Gelman, 'Bewitchment, Biology, or Both: The Co-existence of Natural and Supernatural Explanatory Frameworks across Development,' *Cognitive Science* 32 (2008): 607–42.

198 **American university students:** Weeks, M., and M. B. Lupfer, 'Religious Attributions and Proximity of Influence: An Investigation of Direct Interventions and Distal Explanations,' *Journal for the Scientific Study of Religion* 39 (2000): 348–62.

199 **creation myth:** Corey, M. A., *God and the New Cosmology: The Anthropic Design Argument* (Lanham, MD: Rowman and Littlefield Publishers, 1993); http://xkcd.com/10.

199 **'a globe or a clock':** Cicero, M. T., *The Nature of the Gods*, H. C. P. McGregor, trans. (London: Penguin, 1972; original work published 45 BC), 163.

199 **'We are entirely accustomed':** Dawkins, R., *The Blind Watchmaker: Why the Evidence of Evolution Reveals a Universe without Design* (New York: Norton, 1996; original work published 1987), xvi.

200 **polled by Gallup:** Gallup, 'Four in 10 Americans Believe in Strict Creationism' (December 17, 2010), retrieved from http://www.gallup.com/poll/145286/Four-Americans-Believe-Strict-Creationism.aspx.

200 **'It is almost as if':** Dawkins, *The Blind Watchmaker*, xv.

200 **'artificialists':** Piaget, J., *The Child's Conception of the World*, J. Tomlinson and A. Tomlinson, trans. (Lanham, MD: Rowman and Littlefield Publishers, 2007; original work published 1929).

200 **three in four pre-school children:** Kelemen, D., 'The Scope of Teleological Thinking in Preschool Children,' *Cognition* 70 (1999): 241–72.

200 **ten-year-olds:** Kelemen, D., 'Why Are Rocks Pointy?: Children's Preference for Teleological Explanations of the Natural World,' *Developmental Psychology* 35 (1999): 1440–53.

200 **'intuitive theists':** Kelemen, D., 'Are Children 'Intuitive Theists'?: Reasoning about Purpose and Design in Nature,' *Psychological Science* 15 (2004): 295–301.

200 **adults . . . teleology:** Kelemen, D., and E. Rosset, 'The Human Function Compunction: Teleological Explanation in Adults,' *Cognition* 111 (2009): 138–43.

202 **surprising and significant event:** Weiner, B. ''Spontaneous' Causal Thinking,' *Psychological Bulletin* 97 (1985): 74–84.

202 **car crash:** Lehman, D. R., C. B. Wortman, and A. F. Williams, 'Long-Term Effects of Losing a Spouse or Child in a Motor Vehicle Crash,' *Journal of Personality and Social Psychology* 52 (1985): 218–31.

204 **Even atheists:** Heywood, B. T., ''Meant to Be': How Religious Beliefs, Cultural Religiosity, and Impaired Theory of Mind Affect the Implicit Bias to Think Teleologically' (Queen's Univ. Belfast: unpublished doctoral dissertation, 2010).

205 **V-1 Buzz Bombs:** Gilovich, T., *How We Know What Isn't So: The Fallibility of Human Reason in Everyday Life* (New York: Free Press, 1991); Saunders, H. S. G., 'The Flying Bomb,' *LIFE* 17 (November 20, 1944): 90–99.

205 **British statistician:** Clarke, R. D., 'An Application of the Poisson Distribution,' *Journal of the Institute of Actuaries* 72 (1946): 481.

205 *patternicity:* Shermer, M., 'Patternicity: Finding Meaningful Patterns in Meaningless Noise,' *Scientific American* (November 25, 2008), retrieved from http://www.scientificamerican.com/article.cfm?id=patternicity-finding-meaningful-patterns.

206 **shuffle feature:** Levy, S., *The Perfect Thing: How the iPod Shuffles Commerce, Culture, and Coolness* (New York: Simon and Schuster, 2006).

206 **bad at recognizing and producing randomness:** Nickerson, R. S., 'The Production and Perception of Randomness,' *Psychological Review* 109 (2002): 330–57.

206 **hot hand:** Caruso, E. M., A. Waytz, and N. Epley, 'The Intentional Mind and the Hot Hand: Perceiving Intentions Makes Streaks Seem Likely to Continue,' *Cognition* 116 (2010): 149–53; Gilovich, T., R. Vallone, and A. Tversky, 'The Hot Hand in Basketball: On the Misperception of Random Sequences,' *Cognitive Psychology* 17 (1985): 295–314.

207 **'by at least 12 months':** Newman, G. E., et al., 'Early Understandings of the Link between Agents and Order,' *Proceedings of the National Academy of Sciences* 107 (2010): 17144.

207 **'I got chills':** Taylor, L. C., 'Disney World Photo Captures Couple Together 15 Years before They Met,' *Toronto Star* (June 10, 2010), retrieved from http://www.thestar.com/living/article/821688--disney-world-photo-captures-couple-together-15-years-before-they-met.

208 *synchronicity:* Beitman, B. D. (ed.), 'Synchronicity, Weird Coincidences, and Psychotherapy' [special issue], *Psychiatric Annals* 39(5) (2010).
208 **Mark David Chapman:** Jones, J., *Let Me Take You Down: Inside the Mind of Mark David Chapman, the Man Who Killed John Lennon* (New York: Villard Books, 1992).
210 **'I was glad':** King, R., 'A Match Made at Disney,' *Daily Mail* (June 11, 2010), retrieved from http://www.dailymail.co.uk/news/worldnews/article-1285238/Engaged-couple -discover-paths-crossed-Disney-World-toddlers.html.
210 **'Every single thing':** Brundage, B. E., 'First Person Account: What I Wanted to Know but Was Afraid to Ask,' *Schizophrenia Bulletin* 9 (1983): 584.
210 **images and strings of letters:** Krummenacher, P., et al., 'Dopamine, Paranormal Belief, and the Detection of Meaningful Stimuli,' *Journal of Cognitive Neuroscience* 22 (2010): 1670–81.
210 **unrelated words:** Gianotti, L. R., et al., 'Associative Processing and Paranormal Belief,' *Psychiatry and Clinical Neurosciences* 55 (2001): 595–603.
211 **'Again and again, adoptive parents':** Howell, S., *The Kinning of Foreigners: Transnational Adoption in Global Perspective* (Berghahn Books, 2006), 73.
212 *mere ownership effect:* Beggan, J. K., 'On the Social Nature of Nonsocial Perception: The Mere Ownership Effect,' *Journal of Personality and Social Psychology* 62 (1992): 229–37.
212 *existence bias:* Eidelman, S., C. S. Crandall, and J. Pattershall, 'The Existence Bias,' *Journal of Personality and Social Psychology* 97 (2009): 765–75.
212 **optimizing processes:** Taylor, S. E., and J. D. Brown, 'Illusion and Well-being: A Social Psychological Perspective on Mental Health,' *Psychological Bulletin* 103 (1988): 193–210.
212 **'Technically speaking':** Gilbert, D. T., et al., 'The Illusion of External Agency,' *Journal of Personality and Social Psychology* 79 (2000): 698.
213 **weather simulations:** Lorenz, E. N., 'Deterministic Nonperiodic Flow,' *Journal of the Atmospheric Sciences* 20 (1963): 130–41.
213 *counterfactual thinking:* Epstude, K., and N. Roese, 'The Functional Theory of Counterfactual Thinking,' *Personality and Social Psychology Review* 12 (2008): 168–92.
214 **might-have-been musings:** Kray L. J., et al., 'From What Might Have Been to What Must Have Been: Counterfactual Thinking Creates Meaning,' *Journal of Personality and Social Psychology* 98 (2010): 106–18.
214 *Bockscar:* The Manhattan Project Heritage Preservation Association, 'The Nagasaki Mission – Timeline' (August 3, 2005), retrieved from http://www .mphpa.org/classic/HISTORY/H-07m1.htm; Rhodes, R., *The Making of the Atomic Bomb* (New York: Simon and Schuster, 1986).
215 **'Two additional runs':** Rhodes, *The Making of the Atomic Bomb*, 740.
215 **German motorcyclist:** Teigen, K. H., 'When a Small Difference Makes a Big Difference: Counterfactual Thinking and Luck,' in D. R. Mandel, D. J. Hilton, and P. Catellani (eds.), *The Psychology of Counterfactual Thinking* (London: Routledge, 2005), 129–46.
215 **Karl Teigen:** For an overview, see ibid.
216 **'the typical lucky person':** Ibid., 133.
217 **feelings of gratitude:** Emmons, R. A., and M. E. McCullough, 'Counting Blessings versus Burdens: Experimental Studies of Gratitude and Subjective Well-being in Daily Life,' *Journal of Personality and Social Psychology* 84 (2003): 377–89.
218 **'narrative psychology':** McAdams, D. P., 'The Psychology of Life Stories,' *Review of General Psychology* 5 (2001): 100–122.
219 **'When people believe their lives':** Kray et al., 'From What Might Have Been to What Must Have Been,' 110.

220 **referential thinking:** King, L. A., and J. A. Hicks, 'Positive Affect, Intuition, and Referential Thinking,' *Personality and Individual Differences* 46 (2009): 719–24.

220 **We credit a supernatural one:** Gray, K., and D. M. Wegner, 'Blaming God for Our Pain: Human Suffering and the Divine Mind,' *Personality and Social Psychology Review* 14 (2010): 7–16.

221 **2004 Indian Ocean tsunami:** English, S., 'Tsunami Was 'God's Punishment,'' *Times* (February 10, 2005), retrieved from http://www.timesonline.co.uk/tol/ news/world/article512563.ece; *Washington Times*, 'Allah Off the Richter Scale' (January 9, 2005), retrieved from http://www.washingtontimes.com/news/2005/ jan/09/20050109-102911-9121r/.

221 **Hurricane Katrina:** G., T., 'Is Katrina God's Punishment for Abortion?' [blog post], retrieved from http://www.salon.com/news/politics/war_room/2005/08/30/ hurricane; Martel, B., 'Storms Payback from God, Nagin Says,' Associated Press (January 17, 2006), retrieved from http://www.washingtonpost.com/wp-dyn/ content/article/2006/01/16/AR2006011600925.html.

221 **tsunami that hit Japan:** Dimiero, B., 'Beck: 'I'm Not Not Saying' God Is Causing Earthquakes' [blog post] (March 14, 2011), retrieved from http://mediamatters. org/blog/201103140010.

221 *negative agency bias:* Morewedge, C. K., 'Negativity Bias in Attribution of External Agency,' *Journal of Experimental Psychology: General* 138 (2009): 535–45.

222 **negative experiences:** Baumeister, R. F., et al., 'Bad Is Stronger Than Good,' *Review of General Psychology* 5 (2001): 323–70.

222 **paralysis victims . . . lottery winners:** Brickman, P., D. Coates, and R. Janoff-Bulman, 'Lottery Winners and Accident Victims: Is Happiness Relative?' *Journal of Personality and Social Psychology* 36 (1978): 917–27; (Janoff-) Bulman, R., and C. B. Wortman, 'Attributions of Blame and Coping in the 'Real World': Severe Accident Victims React to Their Lot,' *Journal of Personality and Social Psychology* 35 (1977): 351–63.

223 **'God may be both':** Gray and Wegner, 'Blaming God for Our Pain,' 12.

223 **'To alcohol!':** Swartzwelder, J. (writer) and B. Anderson (director), 'Homer vs. the 18th Amendment' [television series episode], in M. Groening (creator), *The Simpsons* (Los Angeles: 20th Century Fox Television, March 16, 1997).

223 **people are pretty resilient:** Janoff-Bulman, R., and D. Yopyk, 'Random Outcomes and Valued Commitments: Existential Dilemmas and the Paradox of Meaning,' in J. Greenberg, S. L. Koole, and T. Pyszczynski (eds.), *Handbook of Experimental Existential Psychology* (New York, NY: Guilford Press, 2004), 122–40; Tedeschi, R. G., and L. G. Calhoun, 'The Posttraumatic Growth Inventory: Measuring the Positive Legacy of Trauma,' *Journal of Traumatic Stress* 9 (1996): 455–71.

225 **a loving god's plan:** Pargament, K. I., H. G. Koenig, and L. M. Perez, 'The Many Methods of Religious Coping: Development and Initial Validation of the RCOPE,' *Journal of Clinical Psychology* 56 (2000): 519–43.

225 **Cadillacs and jewellery:** (Janoff-) Bulman and Wortman, 'Attributions of Blame and Coping in the 'Real World.''

225 **'automatic punishments':** Piaget, J., *The Moral Judgment of the Child*, M. Gabain, trans. (New York: Free Press, 1997; original work published 1932), 251.

225 **'nature is a harmonious whole':** Ibid., 256.

226 **university students:** Raman, L., and G. A. Winer, 'Evidence of More Immanent Justice Responding in Adults Than Children: A Challenge to Traditional Developmental Theories,' *British Journal of Developmental Psychology* 22 (2004): 255–74.

226 **highly intuitive:** Callan, M. J., R. Sutton, and C. Dovale, 'When Deserving Translates into Causing: The Effects of Cognitive Load on Immanent Justice Reasoning,' *Journal of Experimental Social Psychology* 46 (2010): 1097–1100.

226 **good fortune:** Callan, M. J., J. H. Ellard, and J. E. Nicol, 'The Belief in a Just World and Immanent Justice Reasoning in Adults,' *Personality and Social Psychology Bulletin* 32 (2006): 1646-58.

226 **Gabriel Vivas:** Associated Press, "'Let Us Pray': Peru Survivors Tell Harrowing Tale' (August 25, 2005), retrieved from http://www.msnbc.msn.com/id/9071653/%5Benter%20URL%5D; Caruso, D. B., 'Friends Say Brooklyn Family that Survived Plane Crash Deserved Miracle,' Associated Press (August 25, 2005), retrieved from LexisNexis.

226 **just world theory:** For a review, see Hafer, C. L., and L. Begue, 'Experimental Research on Just World Theory: Problems, Developments, and Future Challenges,' *Psychological Bulletin* 131 (2005): 128-66.

227 **children as young as three:** Olson, K. R., et al., 'Judgments of the Lucky across Development and Culture,' *Journal of Personality and Social Psychology* 94 (2008): 757-76.

227 **everyday behaviour:** Gaucher, D., et al., 'Compensatory Rationalizations and the Resolution of Everyday Undeserved Outcomes,' *Personality and Social Psychology Bulletin* 36 (2010): 109-18.

227 **underground:** I wonder if the gambler's fallacy plays a role in this expectation.

227 **Germany suffered the worst flood disaster:** Otto, K., et al., 'Posttraumatic Symptoms, Depression, and Anxiety of Flood Victims: The Impact of the Belief in a Just World,' *Personality and Individual Differences* 40 (2006): 1075-84; Socher, M., and G. Bohme-Korn, 'Central European Floods 2002: Lessons Learned in Saxony,' *Journal of Flood Risk Management* 1 (2008): 123-29.

228 **search for patterns:** Whitson, J. A., and A. D. Galinsky, 'Lacking Control Increases Illusory Pattern Perception,' *Science* 322 (2008): 115-17.

229 **connect events to later events:** Lindberg, M. J., and K. D. Markman, 'When What Happens Tomorrow Makes Today Seem Meant to Be: The Meaning Making Function of Counterfactual Thinking' (Ohio Univ.: unpublished doctoral dissertation, 2010).

230 **external control:** For a review, see Kay, A. C., et al., 'Religious Belief as Compensatory Control,' *Personality and Social Psychology Review* 14 (2010): 37-48.

231 **locus of control:** Skinner, E. A., 'A Guide to Constructs of Control,' *Journal of Personality and Social Psychology* 71 (1996): 549-70.

231 **abilities to be relatively fixed:** Dweck, C. S., C. Chiu, and Y. Hong, 'Implicit Theories and Their Role in Judgments and Reactions: A World from Two Perspectives,' *Psychological Inquiry* 6 (1995): 267-85.

231 **cancer screening:** de los Monteros, K. E., and L. C. Gallo, 'The Relevance of Fatalism in the Study of Latina's Cancer Screening Behavior: A Systematic Review of the Literature,' *International Journal of Behavioral Medicine* (in press).

231 **diet or exercise:** Shen, L., C. Condit, and L. Wright, 'The Psychometric Property and Validation of a Fatalism Scale,' *Psychology and Health* 24 (2009): 597-613.

231 **heart attack:** Agarwal, M., and A. K. Dalal, 'Beliefs about the World and Recovery from Myocardial Infarction,' *Journal of Social Psychology* 133 (1993): 385-94.

231 **dangerous driving:** Kouabenan, D. R., 'Beliefs and the Perception of Risks and Accidents,' *Risk Analysis* 18 (1998): 243-52.

231 **seatbelt use:** Colón, I., 'Race, Belief in Destiny, and Seatbelt Use: A Pilot Study,' *American Journal of Public Health* 82 (1992): 875-77.

231 **unprotected sex:** Kalichman, S. C., et al., 'Fatalism, Future Outlook, Current Life Satisfaction, and Risk for Human Immunodeficiency Virus (HIV) Infection among Gay and Bisexual Men,' *Journal of Consulting and Clinical Psychology* 65 (1997): 542-46.

231 **destiny as an agent:** Shaffer, L. S., 'Fatalism as an Animistic Attribution Process,' *Journal of Mind and Behavior* 5 (1984): 351–61.

231 **Some fisherman:** Gill, A., 'All at Sea? The Survival of Superstition,' *History Today* 44 (December 1994): 9–11.

231 **American ports:** Dzugan, J., 'Age, Culture and Genes in Risk Taking,' *NORA Proceedings 2006* (2006): 31–38.

232 **blackjack:** Carlin, B. I., and D. Robinson, 'Fear and Loathing in Las Vegas: Evidence from Blackjack Tables,' *Judgment and Decision Making* 4 (2009): 385–96.

232 **vaccine:** Ritov, I., and J. Baron, 'Reluctance to Vaccinate: Omission Bias and Ambiguity,' *Journal of Behavioral Decision Making* 3 (1990): 263–77.

232 **Guinness World Record:** Brackney, S., *Plan Bee: Everything You Ever Wanted to Know about the Hardest-Working Creatures on the Planet* (New York: Penguin, 2009).

233 **'man is condemned to be free':** Sartre, J.-P., *Existentialism Is a Humanism*, C. Macomber, trans. (New Haven, CT: Yale Univ. Press, 2007; original work published 1946), 29.

233 **choice overload effect:** For an overview, see Scheibehenne, B., R. Greifeneder, and P. M. Todd., 'Can There Ever Be Too Many Options? A Meta-Analytic Review of Choice Overload,' *Journal of Consumer Research* 37 (2010): 409–25.

233 **'tyranny of freedom':** Schwartz, B., 'Self-Determination: The Tyranny of Freedom,' *American Psychologist* 55 (2000): 79–88.

233 **'A good life may require constraints':** Markus, H. R., and B. Schwartz, 'Does Choice Mean Freedom and Well-being?' *Journal of Consumer Research* 37 (2010): 352.

233 **faced with a difficult decision:** Beattie, J., et al., 'Psychological Determinants of Decision Attitude,' *Journal of Behavioral Decision Making* 7 (1994): 129–44.

234 **Alexander wakes up:** Viorst, J., *Alexander and the Terrible, Horrible, No Good, Very Bad Day* (New York: Aladdin, 1987; original work published 1972).

235 **bronze medal:** Medvec, V. H., S. F. Madey, and T. Gilovich, 'When Less Is More: Counterfactual Thinking and Satisfaction among Olympic Athletes,' *Journal of Personality and Social Psychology* 69 (1995): 603–10.

236 **negotiating with fate:** Phrase inspired by Au, E. W. M., 'Negotiable Fate: Potential Antecedents and Possible Consequences' (Univ. of Illinois at Urbana-Champaign: unpublished doctoral dissertation, 2008).

238 **'Wisdom is tested':** Birren, J. E., and L. M. Fisher, 'The Elements of Wisdom: Overview and Integration,' in R. J. Sternberg (ed.), *Wisdom: Its Nature, Origins, and Development* (Cambridge, UK: Cambridge Univ. Press, 1990), 324.

238 **make peace:** Morling, B., and S. Evered, 'Secondary Control Reviewed and Defined,' *Psychological Bulletin* 132 (2006): 269–96.

Epilogue: The World Is Sacred

240 **sacredness:** Belk, R. W., M. Wallendorf, and J. F. Sherry Jr., 'The Sacred and the Profane in Consumer Behavior: Theodicy on the Odyssey,' *Journal of Consumer Research* 16 (1989): 1–37; Durkheim, E., *The Elementary Forms of Religious Life*, K. E. Fields, trans. (New York: Free Press, 1995; original work published 1915); Tetlock, P. E., 'Thinking the Unthinkable: Sacred Values and Taboo Cognitions,' *Trends in Cognitive Sciences* 7 (2003): 320–24.

240 **'the value of nothing':** Wilde, O., *Lady Windermere's Fan: A Play about a Good Woman* (London: The Bodley Head, 1893), 95.

241 **sanctification:** For an overview, see Pargament, K. I., and A. M. Mahoney, 'Sacred Matters: Sanctification as a Vital Topic for the Psychology of Religion,' *International Journal for the Psychology of Religion* 15 (2005): 179–99.

241 **mindful meditation:** Goldstein, E. D., 'Sacred Moments: Implications on Well-being and Stress,' *Journal of Clinical Psychology* 63 (2007): 1001–19.

242 **perception of sacredness:** Kesebir, P., C.-Y. Chiu, and T. Pyszczynski, *The Sacred: An Existential Anxiety Buffer* (manuscript submitted for publication, 2010).

243 **'Being human always points':** Frankl, V. E., *Man's Search for Meaning* (New York: Pocket Books, 1984; original work published 1946), 133.

243 **'Events that lack':** Baumeister, R. F., *Meanings of Life* (New York: Guilford Press, 1991), 61.

243 *The Karate Kid:* Weintraub, J. (producer), and J. G. Avildsen (director), *The Karate Kid* [motion picture] (United States: Columbia Pictures, 1984).

244 **'In accepting [the] challenge':** Frankl, *Man's Search for Meaning*, 137.

244 *flow:* Csikszentmihalyi, M., 'If We Are So Rich, Why Aren't We Happy?' *American Psychologist* 54 (1999): 821–27.

245 **'He who has a *why*':** Cited in Frankl, *Man's Search for Meaning*, 97.

245 **rote activities:** Hsee, C. K., A. X. Yang, and L. Wang, 'Idleness Aversion and the Need for Justifiable Busyness,' *Psychological Science* 21 (2010): 926–30.

245 **positive affect:** King et al., 'Ghosts, UFOs, and Magic: Positive Affect and the Experiential System,' *Journal of Personality and Social Psychology* 92 (2007): 905–19; King, L. A., et al., 'Positive Affect and the Experience of Meaning in Life,' *Journal of Personality and Social Psychology* 90 (2006): 179–96.

245 **'every human experience':** Eliade, M., *The Sacred and the Profane: The Nature of Religion*, W. R. Trask, trans. (New York: Harcourt, 1959; original work published 1957), 171.

246 **song 'Miracles':** Posse, I. C., and M. E. Clark, 'Miracles,' on *Bang! Pow! Boom!* [album] (Farmington Hills, MI: Psychopathic, 2009).

246 **'Magical Mysteries':** Michaels, L. (producer), *Saturday Night Live* [television broadcast] (New York: NBC, April 17, 2010).

246 **'A giraffe may not':** Richards, J., 'Violent J of Insane Clown Posse Explains the Remarkable Song 'Miracles'' [blog post] (April 27, 2010), retrieved from http://nymag.com/daily/entertainment/2010/04/violent_j_of_insane_clown_poss.html.

246 **'Garbage magnet':** Spitzer, J. (writer) and D. Holland (director), 'Costume Contest' [television series episode], in G. Daniels (producer), *The Office* (New York: NBC, October 28, 2010).

247 **'*committed* to living':** Janoff-Bulman, R., and D. Yopyk, 'Random Outcomes and Valued Commitments: Existential Dilemmas and the Paradox of Meaning,' in J. Greenberg, S. L. Koole, and T. Pyszczynski (eds.), *Handbook of Experimental Existential Psychology* (New York, NY: Guilford Press, 2004), 130.

247 **'Once our hopeless questioning':** Singer, I., *Meaning in Life: The Creation of Value* (Cambridge, MA: MIT Press, 2010; original work published 1992), 80–81.

247 **'Poets say science':** Feynman, R. P., R. B. Leighton, and M. L. Sands, *The Feynman Lectures on Physics*, vol. 1 (Reading, MA: Addison-Wesley, 1963), 3–6.

FURTHER READING AND
SELECTED BIBLIOGRAPHY

Introduction: We're All Believers

Haselton, M. G., et al. 'Adaptive Rationality: An Evolutionary Perspective on Cognitive Bias.' *Social Cognition* 27 (2009): 733–63.

Jahoda, G. *The Psychology of Superstition*. New York: Penguin, 1969.

Lindeman, M., and K. Aarnio. 'Superstitious, Magical, and Paranormal Beliefs: An Integrative Model.' *Journal of Research in Personality* 41 (2007): 731–44.

Malinowski, B. *Magic, Science, Religion and Other Essays*. Glencoe, IL: Free Press, 1948.

Mauss, M. *A General Theory of Magic*. Translated by R. Brain. New York: Routledge, 2001; original work published 1902.

Rosengren, K. S., C. N. Johnson, and P. L. Harris, eds. *Imagining the Impossible: Magical, Scientific, and Religious Thinking in Children*. New York: Cambridge Univ. Press, 2000.

Tambiah, S. J. *Magic, Science, Religion, and the Scope of Rationality*. Cambridge: Cambridge Univ. Press, 1990.

Zusne, L., and W. H. Jones. *Anomalistic Psychology: A Study of Magical Thinking*. 2nd ed. Hillsdale, NJ: Lawrence Erlbaum Associates, 1989; original work published 1982.

Chapter 1: Objects Carry Essences

Frazer, J. G. *The Golden Bough: A Study in Magic and Religion*. New York: Touchstone, 1995; original work published 1890.

Frazier, B. N., et al. 'Picasso Paintings, Moon Rocks, and Hand-Written Beatles Lyrics: Adults' Evaluations of Authentic Objects.' *Journal of Cognition and Culture* 9 (2009): 1–14.

*Hood, B. M. *SuperSense: Why We Believe in the Unbelievable*. San Francisco: Harper-One, 2009.

Hood, B. M., and P. Bloom. 'Children Prefer Certain Original Objects over Perfect Duplicates.' *Cognition* 106 (2008): 455–62.

Nemeroff, C., and P. Rozin. 'The Contagion Concept in Adult Thinking in the United States: Transmission of Germs and of Interpersonal Influence.' *Ethos* 22 (1994): 158–86.

Chapter 2: Symbols Have Power

Gilovich, T., and K. Savitsky. 'Like Goes with Like: The Role of Representativeness in Erroneous and Pseudo-Scientific Beliefs.' In T. Gilovich, D. W. Griffin, and D. Kahneman, eds. *Heuristics and Biases: The Psychology of Intuitive Judgment*. Cambridge, UK: Cambridge Univ. Press, 2002: 617–24.

Hood, B. M., et al. 'Implicit Voodoo: Electrodermal Activity Reveals a Susceptibility to Sympathetic Magic.' *Journal of Cognition and Culture* 10 (2010): 391–99.

Piaget, J. *The Child's Conception of the World*. Translated by J. Tomlinson and A. Tomlinson. Lanham, MD: Rowman and Littlefield Publishers, 2007; original work published 1929.

Rozin, P., L. Millman, and C. Nemeroff. 'Operation of the Laws of Sympathetic Magic in Disgust and Other Domains.' *Journal of Personality and Social Psychology* 50 (1986): 703–12.

Tylor, E. B., *Primitive Culture: Researches into the Development of Mythology, Philosophy, Religion, Language, Art, and Custom*. 6th ed. Vol. 1. London: John Murray, 1920; original work published 1874.

Chapter 3: Actions Have Distant Consequences

Damisch, L., B. Stoberock, and T. Mussweiler. 'Keep Your Fingers Crossed!: How Superstition Improves Performance.' *Psychological Science* 21 (2010): 1014–20.

Keinan, G. 'The Effects of Stress and Desire for Control on Superstitious Behavior.' *Personality and Social Psychology Bulletin* 28 (2002): 102–8.

Langer, E. 'The Illusion of Control.' *Journal of Personality and Social Psychology* 32 (1975): 311–28.

Risen, J. L., and T. Gilovich. 'Why People Are Reluctant to Tempt Fate.' *Journal of Personality and Social Psychology* 95 (2008): 293–307.

*Vyse, S. A. *Believing in Magic: The Psychology of Superstition*. New York: Oxford Univ. Press, 1997.

Chapter 4: The Mind Knows No Bounds

Baer, J., J. C. Kaufman, and R. F. Baumeister, eds. *Are We Free?: Psychology and Free Will*. New York: Oxford Univ. Press, 2008.

Gilovich, T. *How We Know What Isn't So: The Fallibility of Human Reason in Everyday Life*. New York: Free Press, 1991.

Keltner, D., and J. Haidt. 'Approaching Awe, a Moral, Spiritual, and Aesthetic Emotion.' *Cognition and Emotion* 17 (2003): 297–314.

Neher, A. *The Psychology of Transcendence*. Mineola, NY: Dover, 1990; original work published 1980.

Pronin, E., et al. 'Everyday Magical Powers: The Role of Apparent Mental Causation in the Overestimation of Personal Influence.' *Journal of Personality and Social Psychology* 91 (2006): 218–31.

Chapter 5: The Soul Lives On

Becker, E. *The Denial of Death*. New York: Free Press, 1973.

*Bering, J. M. *The Belief Instinct: The Psychology of Souls, Destiny, and the Meaning of Life*. New York: Norton, 2011.

Bering, J. M. 'The Folk Psychology of Souls,' plus peer commentary. *Behavioral and Brain Sciences* 29 (2006): 453–98.

Bloom, P. *Descartes' Baby: How the Science of Child Development Explains What Makes Us Human*. New York: Basic Books, 2004.

Goldenberg, J. L. 'The Body Stripped Down: An Existential Account of Ambivalence toward the Physical Body.' *Current Directions in Psychological Science* 14 (2005): 224–28.

Greenberg, J., S. L. Koole, and T. Pyszczynski, eds. *Handbook of Experimental Existential Psychology*. New York: Guilford Press, 2004.

Chapter 6: The World Is Alive

Boyer, P. *Religion Explained: The Evolutionary Origins of Religious Thought*. New York: Basic Books, 2001.

Dennett, D. C. *The Intentional Stance*. Cambridge, MA: MIT Press, 1987.

Epley, N., A. Waytz, and J. T. Cacioppo. 'On Seeing Human: A Three-Factor Theory of Anthropomorphism.' *Psychological Review* 114 (2007): 864–86.

Guthrie, S. E. *Faces in the Clouds: A New Theory of Religion*. New York: Oxford Univ. Press, 1993.

Hume, D. *An Enquiry concerning Human Understanding*. New York: Oxford Univ. Press, 2008; original work published 1748.

Chapter 7: Everything Happens for a Reason

(Janoff-) Bulman, R., and C. B. Wortman. 'Attributions of Blame and Coping in the 'Real World': Severe Accident Victims React to Their Lot.' *Journal of Personality and Social Psychology* 35 (1977): 351–63.

Gray, K., and D. M. Wegner. 'Blaming God for Our Pain: Human Suffering and the Divine Mind.' *Personality and Social Psychology Review* 14 (2010): 7–16.

Hafer, C. L., and L. Begue. 'Experimental Research on Just World Theory: Problems, Developments, and Future Challenges.' *Psychological Bulletin* 131 (2005): 128–66.

Kay, A. C., et al. 'Religious Belief as Compensatory Control.' *Personality and Social Psychology Review* 14 (2010): 37–48.

Kelemen, D., and E. Rosset. 'The Human Function Compunction: Teleological Explanation in Adults.' *Cognition* 111 (2009): 138–43.

Kray, L. J., et al. 'From What Might Have Been to What Must Have Been: Counterfactual Thinking Creates Meaning.' *Journal of Personality and Social Psychology* 98 (2010): 106–18.

Sartre, J.-P. *Existentialism Is a Humanism*. Translated by C. Macomber. New Haven, CT: Yale Univ. Press, 2007; original work published 1946.

Shaffer, L. S. 'Fatalism as an Animistic Attribution Process.' *Journal of Mind and Behavior* 5 (1984): 351–61.

Teigen, K. H. 'When a Small Difference Makes a Big Difference: Counterfactual Thinking and Luck.' In D. R. Mandel, D. J. Hilton, and P. Catellani, eds. *The Psychology of Counterfactual Thinking*. London: Routledge, 2005: 129–46.

Epilogue: The World Is Sacred

Baumeister, R. F. *Meanings of Life*. New York: Guilford Press, 1991.

Belk, R. W., M. Wallendorf, and J. F. Sherry Jr. 'The Sacred and the Profane in Consumer Behavior: Theodicy on the Odyssey.' *Journal of Consumer Research* 16 (1989): 1–37.

Durkheim, E. *The Elementary Forms of Religious Life*. Translated by K. E. Fields. New York: Free Press, 1995; original work published 1915.

Eliade, M. *The Sacred and the Profane: The Nature of Religion*. Translated by W. R. Trask. New York: Harcourt, 1959; original work published 1957.

Frankl, V. E. *Man's Search for Meaning*. New York: Pocket Books, 1984; original work published 1946.

Weintraub, J. (producer), and J. G. Avildsen (director). *The Karate Kid* [motion picture]. United States: Columbia Pictures, 1984.

*Together, these three works offer the most recent, accessible, and extended discussions of magical thinking.

INDEX